Hermes Bound

The Policy and Technology of Telecommunications

by Clare D. McGillem and William P. McLauchlan

Science and Society:
A Purdue University Series in Science,
Technology, and Human Values

Hermes Bound

Science and Society:
A Purdue University Series in Science,
Technology, and Human Values

Leon E. Trachtman, General Editor

Volume 2

Hermes Bound

The Policy and Technology
of Telecommunications

by Clare D. McGillem and
William P. McLauchlan

Purdue University
West Lafayette, Indiana
1978

Any opinions, findings, conclusions, or recommendations expressed herein are those of the
authors and do not necessarily reflect the views
of the National Endowment for the Humanities.

Library of Congress Card Catalog Number 77-86267
Printed in the United States of America

Contents

v

Preface

This book began as a study of the impact of telecommunications on society. As the study progressed
it became apparent that in addition to the widely
discussed impacts of telecommunications on social
behavior, business operations, and cultural mores,
society has an equally profound impact on the
development and application of the technology of
telecommunications. This latter phenomenon is
little appreciated by those not directly involved
with policy decisions in the telecommunications
field, and its significance is not always readily
seen when viewed as a series of isolated occur-
rences. However, when the overall picture of the
regulatory, legislative, and judicial control of
telecommunications development is examined clearly,
substantive and far-reaching decisions are being
made in the political arena that determine in
large measure the nature, kind, and timing of
many telecommunications related developments.

Chapter 1 discusses the nature and scope of
the telecommunications revolution that is occurring in the United States and the whole world.
It also briefly considers societal impacts and the
societal controls of telecommunications development.

In order to appreciate the magnitude and significance of the controls society exercises on
telecommunications development, it is necessary
to understand the technological, political, and

ix

economic factors underlying this industry and its
developmental environment. Chapters 2 and 3 introduce the technology of telecommunications, the
political arena and its various actors, and the
economic principles that govern operation of the
telecommunications industry.

Chapters 4 and 5 discuss in detail examples of
the interplay of the technological, political, and
economic factors affecting the introduction or
suppression of new services and technological developments in radio and television broadcasting
and in telephony. Each chapter provides extensive
references to original sources and to additional
related material. The Appendix includes excerpts
from selected legal documents that have played
important roles in telecommunications development.
These excerpts were included to give those readers
not having easy access to such documents an opportunity to study the original materials.

Chapter 6 outlines the general parameters of
technological impact on society. While this book
is focused on the impact of society on technological development, it is worthwhile to provide some
perspective on the other side of the technological-
society relationship. Chapter 6 attempts to make
a set of general statements about the aspects of
society which might reflect the kinds of changes
that telecommunications technology will bring in
years to come.

Chapter 7 focuses on some possible means of
changing or determining the development of technology in this field in the future. Several basic
questions need to be addressed by policy makers
and members of the public, if future development
is going to progress in an understood and systematic fashion. This last chapter examines these
questions. Possibly, the answers to these problems will not emerge from the discussion in
Chapter 7, but government officials and leaders
in the telecommunications industry might consider

suggestions made there as they make future decisions.

   Hermes Bound is not intended primarily as a
textbook although it might be so used in an inter-
disciplinary course or seminar in public policy
and technological development. Rather, the book
is intended to be a detailed overview of the
forces controlling technological development in
telecommunications. As such it should be useful
as a reference and source book for students in
political science, economics, public policy, engineering, communications, and government. No
expertise is assumed on the part of the reader in
any of the areas covered. For those who are
familiar with part or all of the background material contained in Chapters 2 and 3, it would be
appropriate to skip immediately to the later
chapters which are essentially self-contained.

Acknowledgments

The authors would like to express their appreciation to the National Endowment for the Humanities
(grant 0079-65-13505) for their generous support of
the seminars offered under the Science and Culture
Program at Purdue University. Leon E. Trachtman,
associate dean of the School of Humanities, Social
Science, and Education, the recipient of the grant,
provided a great deal of guidance and support
throughout this project, and we appreciate his
contribution. The Purdue Research Foundation also
provided assistance from time to time during the
preparation of this manuscript. Geri Becker typed
the entire manuscript several times, and her prompt
and accurate work made this book much easier to
write than might have been.

Chapter 1

Telecommunications Technology and Society: An Overview

The relationships between technology and society
are complex and multifaceted. One can find literally thousands of empirical studies of these
relationships.(1,2) Much of the emphasis in such
studies is on the impact of technological development on society. However, an equally important,
but much less studied area, is the manner and degree of control of technological development by
society itself. The technology of telecommunications is an example in which the impact of the
technology on society has been profound and is
clearly evident throughout the world. The ways
in which society influences development and application of this technology are much more subtle
and diffuse but nevertheless are also profoundly
important. It is toward a more complete understanding of the relationship between society and
telecommunications technology that this book is
aimed. It will examine the technology itself
along with various political, economic, and social factors that have influenced its development
and utilization.

The early chapters of this book examine the
kinds of influences which economics, politics, and
technical constraints have on the development of
telecommunications. The later chapters explore
how specific decisions affecting telecommunications development were arrived at in the past and
how these decisions altered the development proc-

1

ess. The primary focus of this portion of the
book is on decisions made by public officials, by
people in the telecommunications industry, and by
the consuming public which have significantly
shaped the development of telecommunications in
this country.

Whether or not the forces controlling telecommunications development at any moment are in
proper balance to promote the common good is difficult to assess. By looking in detail at how
these forces have led us to our present development, hopefully, some guidance can be provided
for improving upon or at least preserving the
successful evolution that has occurred in this
field over the past century.

Two basic categories of telecommunications
services and their associated technologies encompass most of the material discussed in the following chapters. The first is telephony and its
subsidiary developments. The second is broadcasting of programs aimed at the commercial attraction of mass audiences. In order to acquaint the
reader with some of the background underlying
these areas a brief introduction to each will be
presented before proceeding with the more detailed
considerations of how they came to their present
development and where they are going.

Telephone Communications

Technologically, the invention of the telephone
was a logical extension of the state of the art
as it existed in the second half of the nineteenth
century. If Alexander Graham Bell had not pat-
ented the telephone in February 1876, someone else
would have within a short time. In fact, there
was considerable litigation over the true inventor
of the telephone before the Supreme Court, in a
dividend vote, upheld Bell's patent. (3) Over the
years improvement in telephone service has con-

tinued as a result of evolution of the technological foundations upon which telephony is based.
From time-to-time major breakthroughs have greatly
enhanced the performance attainable. Important
developments in telephony are listed in Table 1-1
which also shows the nature of the impact that the
development produced. The abbreviated listing in
Table 1-1 omits some highly significant developments. However, it does illustrate some important
aspects of the technological development process.

Mostly, the evolution of telephone technology
has been regular and continuous. At times, however, major technological breakthroughs have led
to quantum jumps in service, performance, or economy. Table 1-1 lists several such developments
that produced major changes in the rate of development and implementation of telephony. The first
was the development of the vacuum triode. Since
its earliest days the telephone was plagued by
the problem of loss of signal intensity as the
distance between the transmitter and receiver increased. This led to serious doubts about the
possibility of long distance communication over
telephone circuits. Invention of the vacuum triode solved this problem by making amplification of
weak signals possible. Thus it was possible to
place repeaters along a transmission line to am-
plify the signals and thereby compensate for the
natural attenuation of the system. The vacuum
triode ushered in a whole era in telecommunications that lasted for half a century. Its effects
extended far beyond telephony into radio, television, and computers and led to the development
of electronics as a major industry of the mid-
twentieth century.

Interestingly, the second revolutionary development in Table 1-1 is the transistor, which was
a precursor of the technology that is replacing
vacuum tubes in telecommunications and a variety
of other fields. While remarkably effective, the

Table 1-1   Key Developments in Telephony

Date	Development	Impact
1876	Telephone patent (Bell)	Basic concept
1877	Carbon transmitter (Edison)	Great increase in signal level
1878	Telephone exchange	Interconnection of telephones
1889	Step-by-step switching (Strowger)	Automatic telephone switching
1899	Loading coils (Pupin, Campbell)	Made long distance possible
1914*	Vacuum triode (DeForrest, Langmuir)	Amplification of signals
1915	Electric wave filters (Campbell)	Allowed many signals on same wires
1918	Carrier telephony	Practical use of one pair of wires for many signals
1919	Crossbar switch (Palmgren, Betulander)	Improved automatic switching
1921	Submarine cable	Showed feasibility of undersea cable
1936	Coaxial cable transmission	Increased transmission capacity and lower cost
1947	Microwave relay	Reduced cost and increased transmission capacity
1948*	Transistor (Brattain, Bardeen, Shockley)	Miniaturization, reliability and cost reduction
1948	Information theory (Shannon)	Gave precise bounds on attainable performance
1951	Customer direct long distance dialing	Improved service
1956	Transatlantic cable	Reliable overseas telephony
1960	Integrated circuits	Extreme miniaturization and reliability
1965	Satellites	Reduced cost and improved service

*Events viewed as technological breakthroughs.

vacuum tube suffered from three serious drawbacks:
its filament or heater generated much unwanted
heat energy that had to be removed when many tubes
were close together; the failure rate was much
higher than would be desired; and the cost was
relatively high. All of these drawbacks have been
essentially eliminated by the transistor and the
vast array of components that have evolved from
the solid state technology on which the transistor
is based.

Although the transistor was invented in 1947,
not until the 1960's was the real impact of this
technology fully felt. Simultaneous fabrication
and mass production of many components into an
"integrated circuit" led to the technological and
economic feasibility of such things as communications satellites and electronic switching systems.
Further technological development led to large-
scale integrated circuits (LSI) that combine many
complex subsystems onto small "chips" of silicon
or other semiconductor materials. This process
led to rapid development of high performance digital computers at remarkably low cost. This
technology is also ushering in a potential revolution in telecommunication services that probably will continue for decades. The manner in
which this almost incredible signal processing
and control technology is brought to fruition will
profoundly influence world social, political, and
economic structures.

The development of technology and its incorpo-
ration into telecommunications is not a free and
independent operation. In fact, many forces are
at play, which strongly affect technological development and evolution. In the early days of
telephony (1873-93) when it held a patent monopoly, the Bell System vigorously opposed any type
of government intervention or regulation.(4) However, when the patent monopoly ended it became
apparent that the only way to achieve stability

and rationality in the telephone industry was
through some type of government regulation. In
1910 the United States Congress passed legislation
which conferred regulatory authority over interstate telephone companies on the Interstate Commerce Commission (ICC).(4) The ICC retained this
jurisdiction until 1934 when regulatory responsibility was transferred to the newly-formed Federal
Communications Commission (FCC). The telephone
industry now operates as a regulated monopoly in
which rates and often services are controlled by
state and federal agencies.

For most of the time since government regulation began, responsible agencies have protected
the telephone monopoly. In recent years, however,
there have been some departures from this policy.
(5) Particularly in the 1968 *Carterfone* decision,
the FCC ruled that private companies would be allowed to connect "foreign attachments" to the
telephone system provided that they complied with
appropriate precautions specified by AT&T, aimed
at protecting the system from possible damage.(6)
The telephone company vigorously opposed this decision but later it acquiesced. Now, such attachments as automatic answering devices, burglar
alarms, and private exchanges are routinely available.

About the same time as the *Carterfone* decision,
the FCC granted Microwave Communications Incorporated (MCI), a specialized carrier, permission to
operate a microwave data transmission link from
St. Louis to Chicago.(7) This directly infringed
on an area previously considered part of the telephone monopoly. Again, AT&T vigorously opposed
the decision, but it also took aggressive action
on the technological front and developed a data
transmission system that was more than financially
competitive with the MCI system.

Telephone system development in the rest of
the world has been largely through government

owned and operated systems and thus does not
closely parallel that of the United States. Most
systems are technically compatible and today a
telephone subscriber in the United States can
reach 90 percent of all the telephones in the
world. Figure 1-1 shows the growth of telephone
installations in the United States and in the
world. The trend is upward, and telephone usage
will likely continue to expand for many years.
However, the number of telephones do not tell the
whole story. The new services that are technically
feasible and may be introduced in coming years
represent an impact at least as great as that of
the increasing utilization of existing services.

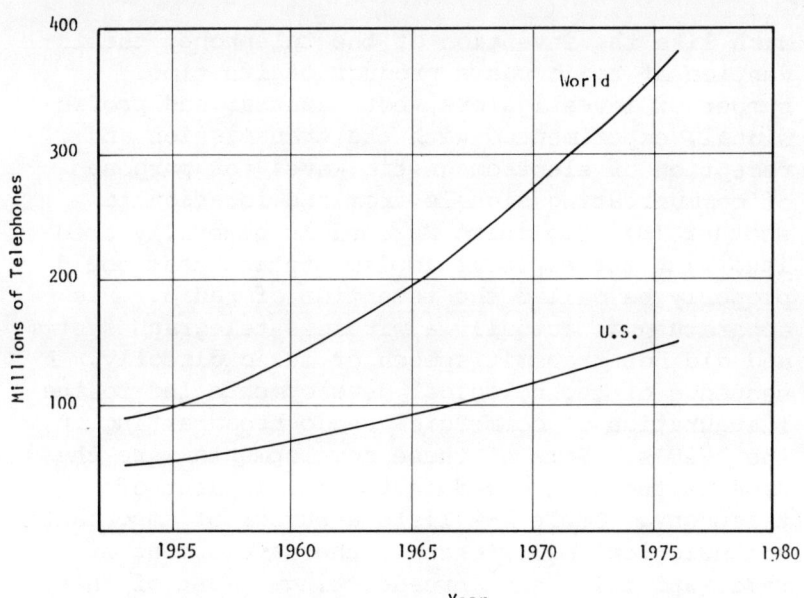

Figure 1-1 Growth of installed telephones
          Source: *Statistical Abstract of U.S.*, 1976

The services and new developments already in
sight include electronic switching systems (ESS)
capable of automatically forwarding your calls to

other numbers, of recording messages, of automatic dialing frequently used numbers, and of many
other conveniences. Domestic satellites will
usher in an era of expanded data communication
services throughout the United States. The use of
optical fibers, thinner than a human hair, to
carry broadband audio, video, and digital communication signals for both intercity transmission and
for distribution to users may herald an era of
greatly expanded telecommunication services in
the home and office, revolutionizing society as
we know it.

Radio and Television Broadcasting

Much like the invention of the telephone, the invention of radio was a product of its time. A
number of investigators, both amateur and professional, experimented with the transmission and
reception of electromagnetic waves for purposes
of communicating signals from one location to
another.(8) Guglielmo Marconi is generally cred-
ited with the explicit implementation that could
properly be called the invention of radio. His
apparatus was actually a wireless telegraph system
and did not transmit speech or music directly. A
sequence of technological developments led to the
inauguration of commercial radio broadcasting in
the 1920's. Some of these developments were the
same as those involved in the development of
telephony. Table 1-2 lists a number of important
technological milestones in the development of
radio and television broadcasting. Most of these
developments were logical extensions of the state
of the art and were absorbed into the broadcasting
industry in a gradual, incremental fashion.

A notable exception to this evolution was the
remarkable and unexpected development of frequency
modulated radio waves (FM). This technological
development represented a major breakthrough be-

cause of its virtual elimination of static and
man-made interference that had plagued radio from
its inception. Despite the superior performance
of radios using this new type of modulation, this
technology has developed much more slowly than

Table 1-2   Milestones in Radio and Television Technology

Date	Event	Impact
1895	Wireless telegraph (Marconi)	Demonstrated feasibility
1903	Transatlantic wireless (Marconi)	First transatlantic tele-communication
1906	Wireless transmission of speech and music (Fessenden)	Established feasibility of radio telephony and radio broadcasting
1906	Audion invented (DeForrest)	Electronic detector
1914*	Vacuum triode (DeForrest, Langmuir)	Electronic amplification
1917	Superhetrodyne receiver (Armstrong)	Great increase in sensitivity
1923	Iconoscope (Zworykin)	Television pickup tube
1932	Television (Zworykin)	Feasibility of TV demonstrated
1933*	Frequency modulation (Armstrong)	Permitted static-free radio
1939	TV broadcasting	Regularly scheduled public TV broadcasts
1949	CATV	Beginning of wired nation
1952	UHF TV	Expanded number of available channels
1953	Color TV	New dimension to TV
1961	Stereo FM	New dimension to radio broadcasting
1973	Quadraphonic FM	New dimension to sound reproduction

*Event considered as technological breakthrough.

was expected and has never displaced AM radio as
the primary broadcasting medium. The reasons for
this are complex and involve economic, political,
and technological forces of various kinds. They
will be discussed in detail in a later chapter.
This suggests that the existence of a superior
technology does not always guarantee that it will
be accepted and used.

The public's acceptance of both radio and television broadcasting was rapid in the United States.
There are almost two radio sets for every person,
a ratio brought about by the very low cost of
manufacturing radios utilizing solid state technology. And, with some 50 million television sets
in the United States, penetration of this potential
market is equally impressive. The low cost and
high reliability of solid state electronic devices
has strongly contributed to this development.
Elsewhere in the world the acceptance of radio and
television has been substantial. In 1974 South
Africa was the largest country without its own
television broadcasting system and it had plans
to inaugurate television broadcasting in 1975 on
a limited scale.(9) Interestingly, the Soviet
Union has the next greatest number of television
sets after the United States, although other
countries exceed the Soviet Union on a per capita
basis.

Societal Impact of
Telecommunications Technology

The size of the telecommunications industry alone
is enough to insure that it will have a major impact on many aspects of society. Even though the
economic impact of telecommunications is enormous,
this is probably not the source of its most significant influence on society. Rather, it is the
increased availability of personal telecommunica-

tion via the telephone. The effects of the mass
media on human behavior has been studied considerably but no definitive conclusions from these
studies have been formed yet.(1,2) At the same
time few persons doubt that a pervasive medium
such as television, which accounts for about the
same number of hours as formal schooling for the
average teenager, will have substantial, long-term
effects.(10) What these effects will be may not
be known for a long time, if ever. Similar un-
answerable questions can be asked about rock-and-
roll and country music that dominate radio broadcasting.

Less obscure effects on social behavior are
directly traceable to telecommunications services.
Among these are the development of new eating
habits such as the TV dinner, the decline and
partial recovery of attendance at movie theaters,
the characteristically protracted telephone conversations of teenagers, the explosion in popu-
larity of professional sports, and the increased
awareness by the general public of minority group
opinions on a vast array of subjects from pollution
to Indian rights.

As more potential services of advanced telecommunications technology are realized, the impacts
on society will be more substantial and farther
reaching. For instance, the conversion to a cashless society would have vast ramifications on the
economic structure and on the ways of doing business and it might eliminate certain types of criminal activity by removing some incentives to com-
mit crime.

Perhaps even more long-term effects would
result from installation of computer-based information terminals in the home. Such terminals
would provide a variety of services such as: shopping catalogues; selectable tutored educational
services; selectable entertainment; specialized
subscriber-oriented news services; automatic bank-

ing; and access to extensive information retrieval
and processing facilities. Such terminals would
have a two-way capability allowing interactive
operations to be carried out via the terminal.
The interactive operations may involve other
people or they may involve computers or specialized automata. Two-way cable television or telephone could well be an inherent part of such a
system. The influence that a telecommunications
system with capabilities such as these could have
on the cultural, economic, and political structure
of a society are apparent.

Impact of Society on Technological Development

The existence of technological potential does not
carry with it any assurance that such potential
will be realized. Or, if it is to be realized,
on what time scale this will occur. Some examples
of this are the slow development of FM radio
broadcasting, the even slower development of UHF
TV broadcasting, the brief introduction and subsequent withdrawal of the Bell System's Picture-
phone® service, and the arrested development of
CATV and pay TV.

What determines the viability of new technology?
In each case different magnitudes and kinds of opposing political, economic, or social forces exist
and ultimate decisions are determined by combinations of these forces. The effect of political
decisions can be of major significance as illustrated by the following example. In the early
1960's government policy was to restrict communication satellite development to a single company
(COMSAT) and this policy was followed during the
entire development of the present international
communication satellite system. However, in the
late 1960's government policy had altered significantly and the development of domestic satellites

was opened to competition. New services and new
technology will likely result from the decision
to open up the field to new participants.

The political decision processes that affect
telecommunications technology development revolve
around the FCC, Congress, the Executive Branch,
the courts, established industry, new entrants,
and other groups such as consumers and private
foundations. Because the economic stakes are high,
there frequently is substantial lobbying by established industry groups when any portion of their
vested interests is threatened by a new entrant.
Such lobbying has generally been effective in that
over the years political decisions have tended to
protect the positions of the established industries. It appears very difficult to obtain political support for innovations if they threaten
established telecommunications industries. This
is particularly true if the industry is part of
the mass media which are molders of public opinion
and which are able to provide subtle positive and
negative inducements to gain congressional support
for their views. This matter is examined in more
detail in a later chapter.

In addition to controlling or influencing telecommunications technology and services through
legislative and regulatory action, there have been
a variety of other attempts ranging from lobbying
by citizen action groups to direct legal action by
dissatisfied users of telecommunications services.
In some cases they have been effective, in others
they have not.

The important point is that there do exist
avenues of control and influence by which society
either directly or indirectly can, and, in fact,
does control the development of new telecommunications technology and services. The role of the
public or consumer in such precedures has been
relatively insignificant over the years but this
is one important factor that seems to be changing.

Many changes in regulatory philosophy and policy
will be required before the consumer's voice is
heard above the cacophony of other voices. Theoretically such changes are possible.

Economics also play a substantial role in determining the development and use of technology.
The future market and regulated economics will
contribute substantially to private and public decisions about the provision of services, the quality of those services, and the pace which innovation takes. Thus, it is important to examine the
structure of economic considerations in order to
appreciate the past developments and to foresee
the likely future with any clarity.

This book explores the technical and political
constraints which shape development in various
telecommunications areas. As a first step it is
necessary to provide the reader with a grasp of
the basic technical factors which are affected by
the economic and political decisions being made.
That is the purpose of Chapter 2. Subsequent
chapters will examine the other factors which influence technological developments, and will explore specific cases in which these factors have
interacted to produce decisions which have controlled specific developments.

Chapter 2

Telecommunications Technology

There have been so many remarkable technological innovations in recent years that many people have come to accept them without any attempt to appreciate their underlying principles or to understand the basic limitations which govern the performance that is attainable. Certainly, obtaining a detailed knowledge of current telecommunication technology would require a great deal of study, far more than the nonspecialist would be willing to invest. Fortunately, however, a few basic concepts are readily understandable and provide an adequate foundation for appreciating many of the most important quantitative and qualitative aspects of telecommunications technology and the equipment utilized for implementing many available services.

This chapter intends to present in understandable form sufficient basic concepts underlying telecommunications theory to allow the nonexpert to understand the terminology of the engineer or scientist and to form his own judgment about the feasibility, desirability, and costs associated with present and future developments in this field. A glossary of terms at the end of the book provides a ready reference to the meaning of these terms when they are encountered throughout the book. (1,2)

Experimental Development
Versus Theoretical Development

Much of the early development of telegraphy, telephony, and radio technology was based on an empirical or experimental approach. Such men as
Morse, Bell, Edison, Marconi, de Forrest, and
others of their era utilized great ingenuity, per-
serverance, and often intuitive insight in developing and perfecting their inventions. The theoretical basis of telecommunications virtually did not
exist. Many of the underlying physical and mathematical laws governing the devices in use were
known but had not been organized into a body of
knowledge that was useful in explaining their operation or in extending their design to more
advanced devices. During the 1920's and 1930's
the theoretical basis of communications expanded
continually. However, not until World War II was
the full power of theoretical analysis brought to
bear on development of the telecommunications technology. This came because of the urgent need to
develop sophisticated radar, sonar, and communication equipment. To meet this need, special
laboratories were formed and many highly capable
physicists, mathematicians, and engineers were
brought into these development programs. Except
for the war, many of these individuals would never
have been engaged in what has become known as communications engineering. Their impact was enormous
both in their wartime accomplishments and in their
long-term effects of placing communications engineering on a much more sound theoretical basis.

Shortly after World War II the analytical basis
of communications science reached a pinnacle with
publication by Claude Shannon of a paper entitled,
"The Mathematical Theory of Communication."(3)   In
this paper Dr. Shannon presented a new, theoretical
way of looking at the transmission of information.

He defined information in a general and useful
manner and showed the limits of communication performance achievable in terms of the speed of
transmission, the characteristics of the signal
being transmitted, and the noise interfering with
that transmission. Although the theory is abstract, many of the results are simple and give
much insight that had previously been only intuitive or empirical in nature. Since then the
theoretical basis of communication engineering has
been continually expanding and today provides a
means for accurately predicting the performance
possible with various systems, for determining how
well a system performs compared with the ultimate
performance possible, and for comparing various
alternate schemes before selecting one for a particular application.

Much experimental and empirical work is still
conducted in communications engineering, particularly in the area of devices and components. This
is because Shannon's analysis did not provide any
practical means for obtaining ultimate performance,
although it showed what that performance was. The
goals of efficient communication are clearly
stated but the paths to their achievement must be
searched out individually. This is one aspect of
communication engineering that has made it fasci-
nating to so many people in the past two decades
and that has contributed to the continual advances
in technology past, present, and future.

Signals and Information

To begin a discussion of telecommunications technology, consider the electric telegraph and how
it transmits information. Figure 2-1 shows an
elementary telegraph system consisting of a battery, a telegraph key, a conductor, and an electromagnet. When the key is closed, the electric circuit is completed from the battery to the electro-

magnet and back through the ground which is also
a good conductor of electricity. With the circuit
closed, current flows through the electromagnet
causing it to pull the lever down. When the key
is released, the circuit is broken, stopping the
current from flowing through the electromagnet,
and, with no current, the electromagnet loses its
pull, and the lever is released and pulled up by
a spring.

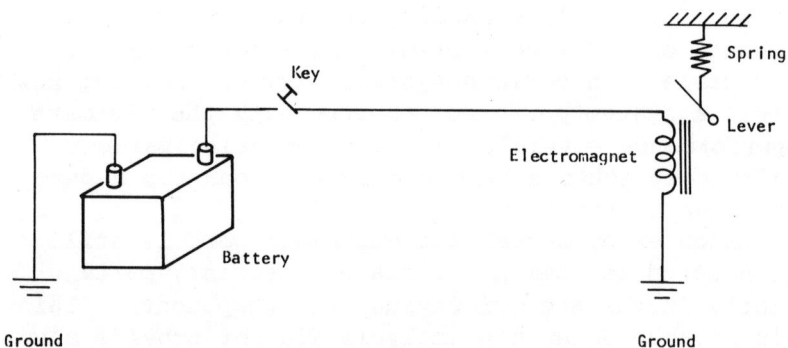

Figure 2-1 Elementary telegraph system

    Signaling is accomplished by varying the
length of time that the key is held down. For
example, in the Morse Code used for radio telegraph transmissions the signals consist of a dot
(key depressed briefly), a dash (key depressed
three times as long as for a dot), and a space
(key up). The complete Morse Code alphabet is
shown in Table 2-1.
    A signaling system such as the electric tele-
graph is called a binary system because it has
essentially two states, open or closed, i.e.,
signal or no signal. Much study has been devoted
to the communication problem as it relates to binary systems. The individual characters of a
binary system are called binary digits and are
usually referred to as *bits*. For example, a system that transmits ten characters each second is

said to have a transmission rate of 10 bits/sec.
This concept of transmission rate is very important for, as will be seen shortly, transmission
of all kinds of information can be measured in
terms of these same units.

Table 2-1 Morse code

A · –	H · · · ·	O – – –	V · · · –
B – · · ·	I · ·	P · – – ·	W · – –
C – · – ·	J · – – –	Q – – · –	X – · · –
D – · ·	K – · –	R · – ·	Y – · – –
E ·	L · – · ·	S · · ·	Z – – · ·
F · · – ·	M – –	T –	
G – – ·	N – ·	U · · –	

More general coding schemes can be used with
binary signaling systems. The most widely used
methods operate at a constant rate of character
transmission. If we consider the two states of
the (binary) system as corresponding to the symbols 0 and 1, then a typical signaling sequence
or message waveform might look as in Figure 2-2.

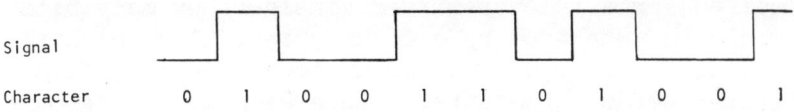

Signal

Character   0   1   0   0   1   1   0   1   0   0   1

Figure 2-2 Binary signal and corresponding characters

Groups of the characters can be taken together to
represent a specific symbol or word. For example,
in the case of teletypewriters the binary characters are taken as groups of five and the various
combinations and permutations of zeros and ones
that can be made in five characters are assigned

specific symbols on the typewriter. Table 2-2
shows what typical five bit code words correspond
to, for some commonly used teletypewriter applications. When a key is pressed, it produces the
five characters shown and when the teletypewriter
receives a particular five bit sequence it causes
the corresponding key to be activated thereby
printing that character on the paper.

Table 2-2 Teletype binary codes

| Code Group | Lower Case | Upper Case for Three Different Codes | | |
	All Codes	Weather Code	AT&T Fraction	Int'l Alphabet No. 2
1000	A	↑	-	-
10011	B	⊕	5/8	?
01110	C	⊖	1/8	:
10010	D	↗	$	who are you?
10000	E	3	3	3
10110	F	→	1/4	not assigned
.	.	.	.	.
.	.	.	.	.
.	.	.	.	.

To get an idea of the rate of transmission required for a teletypewriter consider how many bits
would be generated by a fast typist. Assume that
typing is being done at an average of eighty five-
letter words per minute. Since each letter requires five bits it follows that each five-letter
word requires 25 bits plus five for the space between words. The transmission rate is therefore
80 x 30 = 2400 bits per minute or 2400 ÷ 60 = 40
bits per second. An automatic transmission system that employs a prepunched tape will operate
at speeds up to 150 bits/sec. which is several
times faster than a human operator is able to
type.

Frequency and Bandwidth

Light, sound, radio waves, and telephone signals
are all most conveniently described in terms of
their frequency components. Frequency is the
rate at which a periodic signal repeats itself.
For example, consider sound waves. A violin
string is vibrated by drawing the bow across it.
A heavy, long string will vibrate slowly, while a
light, short string will vibrate more rapidly.
Changing the tension in the string also alters the
frequency of vibration. It is the vibration of
the strings, coupled to the air through the sound-
ing board body of the violin that produces the
sound heard by the ear. The sounds produced by a
violin have frequencies from hundreds to thousands
of cycles per second. The unit of frequency is
the hertz (abbreviated Hz) and one cycle per second corresponds to one hertz. A thousand cycles
per second is one kilohertz (abbreviated kHz) and
a million cycles per second is one megahertz
(MHz).

Human speech is generated by a complex interaction of the vocal chords and various cavities
and structures of the head. All sound waves are
pressure variations that are traveling through the
air, or some other suitable medium, and when they
strike an appropriate sensing device, such as an
ear or a microphone, they set up vibrations that
are detected. The ear is sensitive to sound vibrations in the range of 30 to 20,000 Hz, although
certain ranges are much more important than others
for conveying the information content of speech or
music. Figure 2-3 shows a curve of the relative
loudness to the human ear of sounds having the
same amplitude but different frequencies. The
maximum response is near a frequency of 1 kHz,
which falls off rapidly on either side of this
peak. Actually, the telephone system only carries
speech components whose frequencies lie in the

range of 300 Hz to 3400 Hz. Thus, the bandwidth
(B.W.) of the telephone signal is: 3400 - 300 =
3100 Hz or 3.1 kHz. In an actual system, a bandwidth of 4 kHz is normally allocated to speech
transmission to allow guardspaces to be placed at
the edges of the transmission band. This permits
speech channels to be located adjacent to each
other without interference.

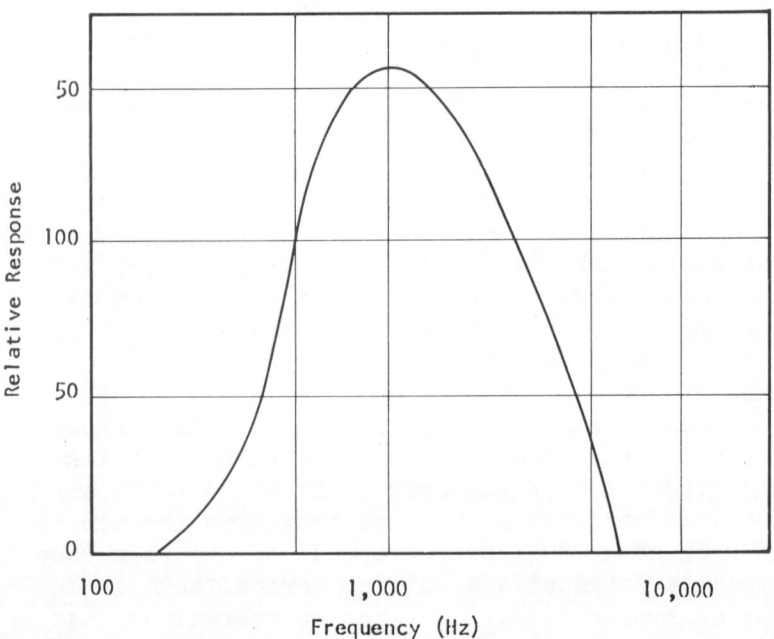

Figure 2-3 Frequency response of ear

    Music sounds tinny when transmitted over a
bandwidth no greater than that of the telephone.
Fidelity improves when the bandwidth is increased
to 5 kHz as in AM radio broadcasting or to 15 kHz
reproduction over the entire bandwidth of the ear
which is 20 to 20,000 Hz. This amounts to a bandwidth that is (20,000 - 20) ÷ 3100 = 6.4 times
greater than that of the telephone. Since the

cost of transmitting signals is a direct function
of their bandwidth, clearly telecommunication of
music is significantly more expensive than that
of speech. The problem is compounded further because good music reproduction requires lower noise
channels than does typical speech communication.
This causes a further increase in the cost of
transmitting music.

Figure 2-4 shows the relation between the frequency scale and various familiar sounds.

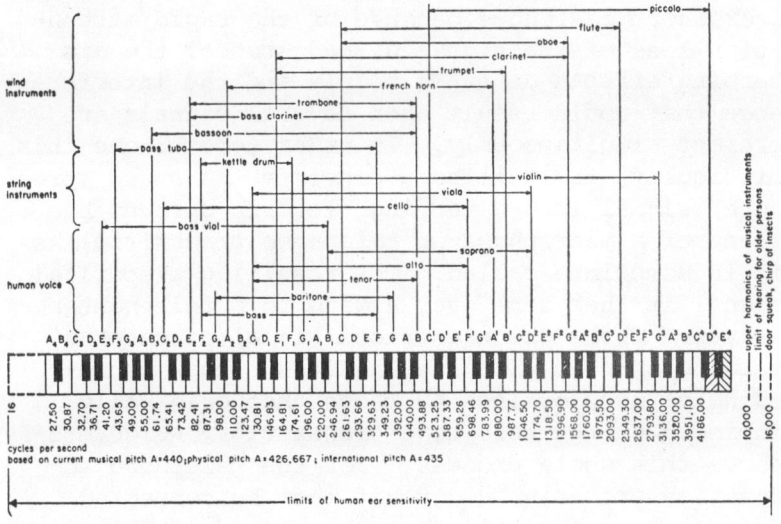

Figure 2-4   Spectra of sounds (From *Reference Data for
Radio Engineers*, Howard W. Sams Co., Inc.,
1975).

The key to successful transmission of signals
corresponding to speech and music is to have sufficient bandwidth in the transmission channel. As
the variations in amplitude of a signal become
more rapid and complex, a greater transmission
bandwidth is required for its faithful reproduction. For this reason, transmission of music requires a much greater bandwidth than does trans-

mission of telegraph signals. It will be seen
shortly that the information bandwidths of 4 kHz
for telephone transmission of speech and 5 or 15
kHz for music are vitally important in determining
how many radio stations can be on the air at one
time or how many telephone conversations can be
handled by a communications satellite.

Telephone Transmission

Communication by sound waves is only feasible
over short distances because of the rapid attenuation loss of the signal with distance, the disturbing effects on other people and the interference that would result when several signals are
present simultaneously. In order to overcome this
difficulty, the telephone converts the sound (pressure) signal into a varying electric current by
means of a microphone or *telephone transmitter* as
it is sometimes called. The resulting electrical
signal is then sent over a wire to a telephone receiver. There the electrical signal actuates a
metal diaphragm which regenerates sounds corresponding more or less to those that originally
impinged on the telephone transmitter. Figure 2-5
shows this whole process. For the telephone to
function properly, a circuit must be connected
from each transmitter to each receiver. This is
called a two-way voice circuit.

In the early days, telephone service was limited to short ranges because of the attenuation
of the signal along the telephone wires. Long
distance communication was measured in tens of
miles. With the advent of the vacuum tube amplifier in 1914, the means for obtaining true long-
distance communications was at hand. By installing amplifiers, or *repeaters* at appropriate
intervals along a telephone line it was possible
to maintain a high signal level and to achieve a

true long distance capability. Telephone service
between New York and San Francisco was inaugurated
in 1915. Transatlantic telephone service using
submarine cable with undersea repeaters was inaugurated in 1956 between Newfoundland and Scotland.

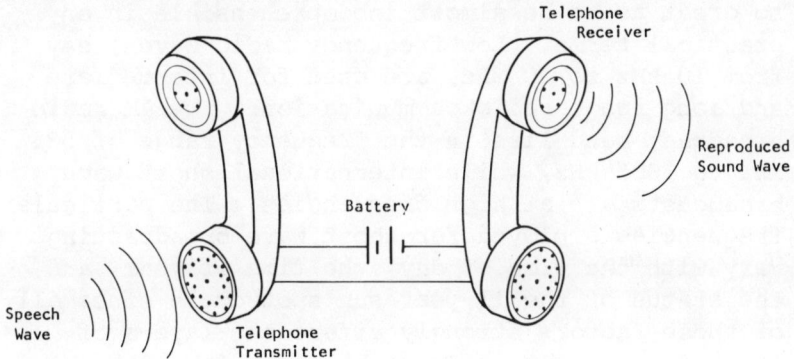

Figure 2-5 Elementary telephone system

Although providing a good communication link
between two telephones is an important and difficult task, one that is equally important and
actually more complex, is that of interconnecting
all of the telephones so that each may call any
other. This is the problem of telephone switching. It has evolved from rudimentary manual interconnection of a few telephones by an operator
in a central office to computer-controlled switching systems that can now interconnect hundreds of
millions of telephones, often in a matter of
seconds.

Radio Waves

Simultaneously with the development of the telephone came the development of radio. It had been
discovered in the late 1800's that there existed
certain kinds of electric waves that could travel

through space without the need for conducting
wires. These waves are called electromagnetic
waves and are manifested as x-rays or light waves
when their frequencies are very high and as radio
waves when their frequencies are relatively low.
The electromagnetic spectrum extends over a range
so great as to be almost incomprehensible in any
practical terms. Low frequency radio waves, say
from 10 kHz to 30 MHz, are used for intermediate
and long range radio communication. The AM radio
broadcast band lies in the frequency range of 535
kHz to 1605 kHz, while international short wave
broadcasts are at high frequencies. The particular
frequencies employed for short wave broadcasting
vary with the time of day, the time of year, and
the status of the 11-year sun spot cycle since all
of these factors strongly affect the layers of
ionized gases surrounding the earth (the ionosphere) which reflect the signals back to earth
permitting long distance communication. The frequencies used for international and other long
range point-to-point communication are in the range
of 3 to 30 MHz. Above about 30 MHz, electromagnetic waves are not reflected back to the earth by
the ionosphere and so communications are limited
to line-of-sight distances between antennas. It
is in this band of frequencies, above 30 MHz, that
television, FM radio, and police and emergency
services lie. The upper frequency limit for this
type of communication is about 1000 MHz or 1 GHz.
Above this frequency radio waves fall into a category of radiation referred to as microwaves. These
frequencies have found wide usage as part of the
microwave relays that interconnect cities and provide for transmission of telephone calls, and
radio and television network broadcasts. Microwave frequencies are also used for the up-and-down
links in satellite communication. Figure 2-6 sum-
marizes the usage of the electromagnetic spectrum.

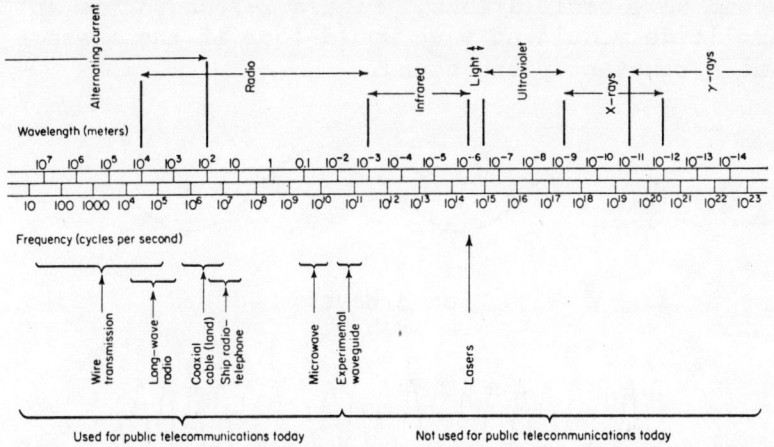

Figure 2-6   The electromagnetic spectrum (From J. Martin, *Telecommunications and the Computer*, Prentice Hall, Inc., 1969).

AM Broadcasting

Radio waves are used for communication by attaching the desired information to the radio wave at the transmitter and then removing the information at the receiver. Thus, the radio wave acts as a carrier. The process of attaching the information to the carrier is called modulation and the process of recovering the information is called demodulation or detection. A common type of modulation is that in which the amplitude (i.e., the magnitude) of the radio wave is caused to vary in the same manner as the sound pressure of the voice or music that is to be transmitted. When the sound is loud, the amplitude is large, and when the sound is soft, the amplitude is small. When the sound varies in pitch, the amplitude of the carrier varies in synchronism with the sound pressure oscillations. The radio wave is itself oscillating at a very high frequency--thousands to millions of times higher in frequency than the

sound wave oscillations. Figure 2-7 shows how an
amplitude modulated wave would look if the message
was a constant pitch sound.

Sound Wave

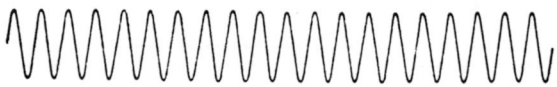

Carrier

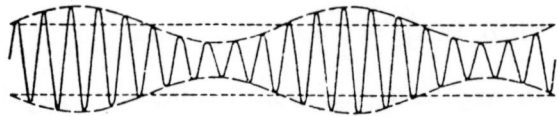

Modulated Carrier

Figure 2-7 Amplitude modulation

   After the radio wave has been modulated, it is
amplified and then radiated into space by an antenna. A receiver extracts the information contained in the modulated waveform. The receiver
performs three functions. First, it separates the
desired carrier from others by a tuning operation
in which the receiver is made sensitive only to
radio waves having the desired frequency. Second,
the receiver removes the carrier and keeps only
the modulation waveform that represents the desired information. This is the detection process.
Third, the receiver amplifies the modulation and
applies it to a loudspeaker where it is reproduced
as sound.

The process of amplitude modulation causes the radio frequency carrier to be spread over a band of frequencies equal to twice the modulation frequency. For AM broadcasting the highest modulation frequency is normally limited to 5,000 Hz or slightly less than twice the bandwidth of a telephone voice circuit. Even so, this means that a single station will occupy a segment of the radio frequency spectrum that is 10 kHz wide, centered on the assigned carrier frequency. If any other radio station broadcasts within the band or near enough so that its modulation components fall in this band, there will be interference between the two stations.

To minimize interference between stations the Federal Communications Commission was given authority to allocate the spectrum by means of licensing procedures. Unfortunately, sufficient space is not available in the radio spectrum to meet all of the demand and less than ideal operations result. This is apparent when tuning over the AM broadcast band at night when the problem of overcrowding is severely aggravated because of the increased transmission distances that result from changes in the earth's ionosphere after sunset. The interference problem can be partially solved by restricting some broadcast stations to daytime broadcasting only, or by requiring reduced power at night. As a further solution to the problem of interference among stations having the same assigned carrier frequencies, 39 clear channels have been established in the western hemisphere on which only one station broadcasts. These channels are widely dispersed geographically and assure interference-free reception in their primary coverage areas. This course of action is only possible to a limited extent because, if the entire broadcast band covering 535 to 1605 kHz were allocated on this basis, there could only be (1605 - 535) ÷ 10 = 107 stations. Actually, there are thousands

of AM stations, and, therefore, interference. The
problem is particularly bad at geographical locations outside the primary coverage area of stations,
because there many signals on the same frequency
may be received simultaneously with more or less
equal power and they interfere with each other.

FM Broadcasting

One of the major disadvantages or limitations of
AM radio is its sensitivity to electrical disturbances of both natural and man-made origin.
Such disturbances manifest themselves as static
and result from such things as lightning, neon
signs, and electric razors. Most natural and man-
made electromagnetic disturbances add to the signal at the receiver and so are detected as a change
in the amplitude and are superimposed on the true
signal contained in the modulation.

In the 1930's Major Edwin H. Armstrong devised
a way to virtually eliminate the susceptibility of
radio signals to static that so severely limits AM
broadcasting. He did this by changing the way in
which he modulated the carrier. Instead of varying
the amplitude of the carrier, Armstrong kept the
amplitude constant and varied the frequency. Figure 2-8 illustrates the frequency modulation of a
carrier. The great improvement in noise performance comes because the receiver can now be made
insensitive to amplitude variations caused by
static. It does this by ignoring the amplitude
variations and looking only for the frequency variations which are virtually unaffected by static.
The technique works effectively and eliminates the
static so prevalent in AM broadcasting.

As we might expect, a price must be paid for
improved performance, and the price in this case
is increased bandwidth. Frequency modulation obtains its improved performance by requiring a
larger segment of the radio spectrum to transmit

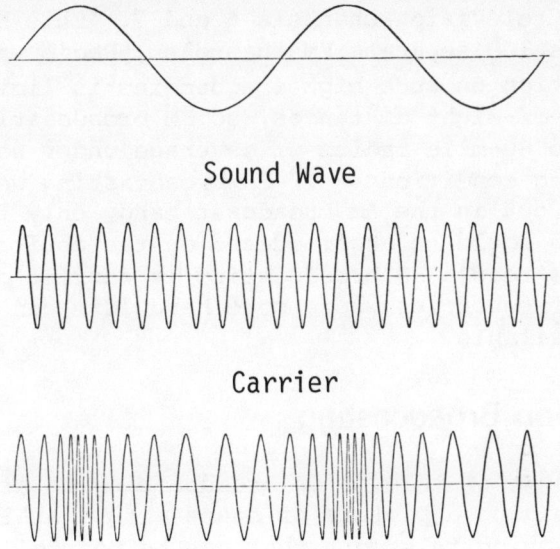

Sound Wave

Carrier

Modulated Carrier

Figure 2-8   Frequency modulation

the same information. In commercial FM broadcasting the ratio of transmission bandwidth to modulating signal bandwidth is about 10:1, compared
with 2:1 for AM broadcasting. Because of the
noise-free performance possible with FM broadcasting, the FCC decided to permit a larger bandwidth
of modulation to be used so that higher quality
music reproduction would be possible. The modulation bandwidth of commercial FM broadcast stations
is 15 kHz which covers the range of most of the
sounds heard by the human ear. The bandwidth of
the transmitted signal is, therefore, 10 x 15 =
150 kHz, which is 15 times greater than that employed in AM broadcasting. In order to minimize
interference between stations on adjacent frequencies, a separation of 200 kHz is used between
channels. The FM broadcast band employed in the
U.S. and Canada is from 88 to 108 MHz, which lies

between television channels 6 and 7. This band
provides 100 separate FM channels. Radio wave
propagation on such high frequencies is limited
to line-of-sight distances, so FM broadcasting has
a 40- to 80-mile radius of coverage under normal
operating conditions. If FM broadcasting were
carried out in the AM broadcast band, only five
stations could be accommodated on the entire band.
For this reason FM broadcasting is carried out at
higher frequencies where more bandwidth can be
made available.

Television Broadcasting

The concept of television broadcasting is simple. Imagine a picture or scene with a small,
light-sensing telescope that can be pointed at
various parts of the scene. The output of the
telescope is an electrical signal proportional to
the intensity of the scene in the direction that
the telescope is pointing. As the telescope is
moved along a horizontal line the signal will vary
as the scene intensity varies along the line.
This electrical signal amplitude-modulates a carrier which is radiated to a receiver where the
wave is demodulated. The information on scene
brightness is sent to a light source that moves in
synchronism with the sensing telescope at the pick-
up location. When the scene brightness increases,
the light intensity increases, and when the scene
brightness decreases, the light intensity decreases. Thus, as the telescope moves over the scene
at the pickup location, the light at the receiver
moving in synchronism with it paints out the same
variations of lightness and darkness as the original scene. By doing this very rapidly (e.g.,
broadcast TV traces the complete picture 30 times
per second) the eye in effect "sees" a continuous
scene without any apparent motion of the scanning
light source.

Figure 2-9 shows a schematic representation of this process. In an actual television system, the moving light on the receiver is an electron beam illuminating the phosphorent screen of a cathode ray tube and the image pickup device is a special television sensor called an image orthocon.

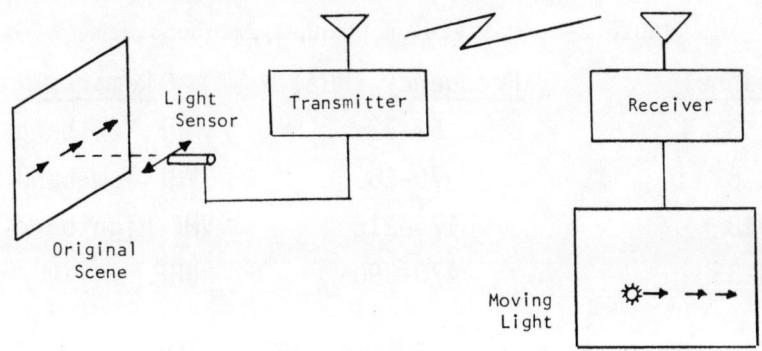

Figure 2-9   Elementary television system

In a typical (American) television system there are 525 lines in the picture and 30 complete frames per second. The horizontal to vertical aspect ratio (i.e., the ratio of width to height) is 4:3. A standard TV picture has approximately 165,000 distinct picture elements (pixels) and when such pictures are reproduced 30 times per second, they occupy a signal bandwidth of 4 MHz. American TV employs a special type of amplitude modulation that utilizes much less bandwidth than conventional AM modulation 1.1:1 as compared to 2:1. When additional spectrum is added for FM transmission of the sound to accompany the picture and a small guard space between channels is included, the total bandwidth requirement is 6 MHz for a single television channel. This is nearly six times the bandwidth of the entire AM broadcast band and the entire FM broadcast band would accommodate only three TV channels. The frequencies utilized for

television broadcasting are shown in Table 2-3.
The noncontiguous spacing of the VHF channels is
not happenstance but intentionally done to minimize certain kinds of interference that can result
when the channels are all adjacent to each other.
(4)

Table 2-3   Television channel frequencies

Channel	Frequency (MHz)	Remarks
2, 3, 4	52-72	VHF low band
5, 6	76-88	VHF low band
7-13	174-216	VHF high band
14-83	470-890	UHF

Video Telephone

The videotelephone as implemented by the Bell
System in its Picturephone® is a miniature television system.   The screen size is 14 cm x 12.7 cm
(5½ in. x 5 in.).   There are 250 lines per frame
and 30 frames per second.   The number of picture
elements in the image is substantially less than
that of commercial TV and also the number of gray
shades is smaller.   Because fewer elements need
be transmitted each second, it requires less transmission bandwidth. The Picturephone® requires only
a 1 MHz bandwidth for satisfactory picture reproduction.   By comparison, this bandwidth would accommodate 250 voice circuits.

Pulse Code Modulation

In recent years a new type of modulation has become very important for a wide variety of communications other than broadcasting.   This new type of
modulation is called pulse code modulation or PCM.

It is of particular interest in a general discussion of telecommunications because it provides a simple method of representing general signals in terms of the elementary form of binary digits or bits. The key concepts required in the understanding of PCM are those of *sampling* and *quantization*. Sampling means measuring the amplitude at equally spaced intervals called the *sampling period*. Quantization is the process of assigning the actual amplitude to the closest one of a predetermined set of (usually) equally spaced amplitude levels. Figure 2-10 illustrates the sampling and quantization of a waveform.

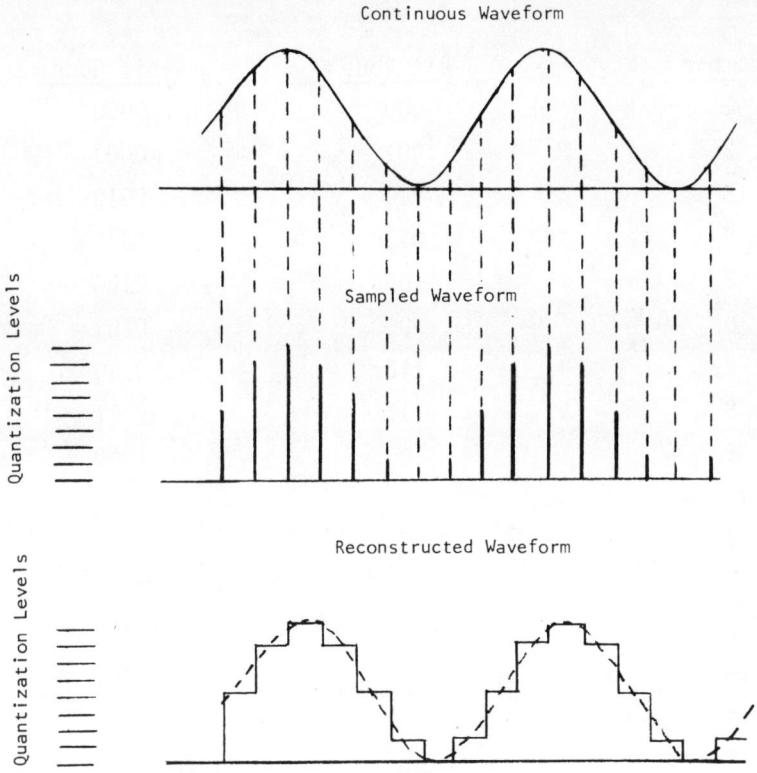

Figure 2-10   Sampling and quantization

Once the sampling and quantization has been done, it is possible to describe the result completely by a sequence of numbers. The first number is the quantum level of the first sample, the second number is the quantum level of the second sample, etc. More importantly, this procedure allows a continuous waveform to be transmitted over a communication channel as a set of discrete numbers as opposed to a continuous or analog waveform. A simple way of transmitting the numbers representing the quantum levels is by means of a binary code similar to that used by a teletypewriter. Examples of 8-level and 16-level binary codes are shown in Table 2-4. The order of the

Table 2-4   Binary codes

Number	3-bit Code	4-bit Code
1	000	0000
2	001	0001
3	010	0010
4	011	0011
5	100	0100
6	101	0101
7	110	0110
8	111	0111
9		1000
10		1001
11		1010
12		1011
13		1100
14		1101
15		1110
16		1111

binary digits determines the number that is represented. Generally, the number of different quantum levels that can be represented by a binary
code with N digits is two raised to the Nth power.
Thus a 4-bit binary code would correspond to 2^4 =
16 levels and a 10-bit code would correspond to
2^{10} = 1024 levels. An illustration of a quantized
waveform is shown in Figure 2-11 along with the
3-bit code sequence that would represent the first
six samples.

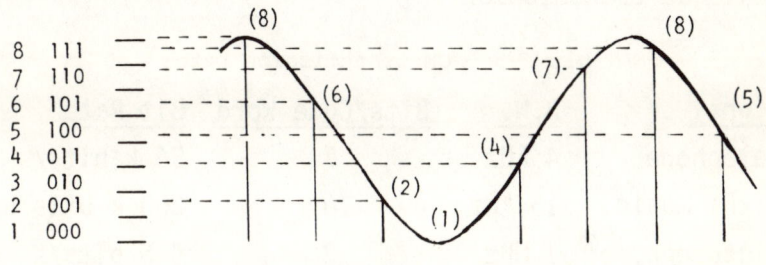

Figure 2-11 PCM signal

The number of quantum levels required depends
on the fidelity with which it is necessary to re-
produce the original waveform. If too few levels
are used, noticeable distortion occurs. If too
many levels are used, no further improvement is
obtained because noise in the transmission intro-
duces distortions larger than the incremental
errors of the quantization.

The rate of transmission required for pulse code
modulation can be obtained by multiplying the number of bits in the code words used to represent the
function levels by the number of samples per second. The sampling rate is always taken as twice

the signal bandwidth to permit undistorted reproduction of the waveform. This can be expressed in
symbols as T = 2WN: where T is the transmission
rate in bits per second; W is the bandwidth of the
signal in Hz; and N is the number of bits in the
code words. The PCM transmission rates required
for transmission of typical telecommunication signals is shown in Table 2-5. It is interesting to
note from Table 2-5 that with PCM the transmission
rate of Picturephone® signals is only 100 times the
rate of a voice channel compared with 250 times for
analogue transmission.

Table 2-5 PCM transmission rates

Signal	B.W.	Bits/Code Word	Bit Rate
Telephone	4 kHz	7	56 k bits/s
Hi-Fi music	15 kHz	10	300 k bits/s
Picturephone®	1 MHz	3	6 M bits/s
Color TV	4.6 MHz	10	92 M bits/s

The primary attractions of PCM are its inherent
insensitivity to noise since only on-off type signals are transmitted and the ease with which the
signal can be exactly regenerated during transmission over long distances. Also, the format of
PCM is closely related to that of computer data and
can be transmitted over the same channels using the
same equipment.

Channel Capacity and Multiplexing

Much effort has been expended over the years in
expanding the capacity of communication systems.
Figure 2-12 shows how this growth has proceeded
over the past 125 years. It is evident from the
figure that channel capacity is increasing at a
phenomenal rate--by a factor of 10 times each 18

years. If this pace continues, and there is every
reason to believe that it will, the capability of
greatly expanded telecommunications services will
be at hand almost immediately.

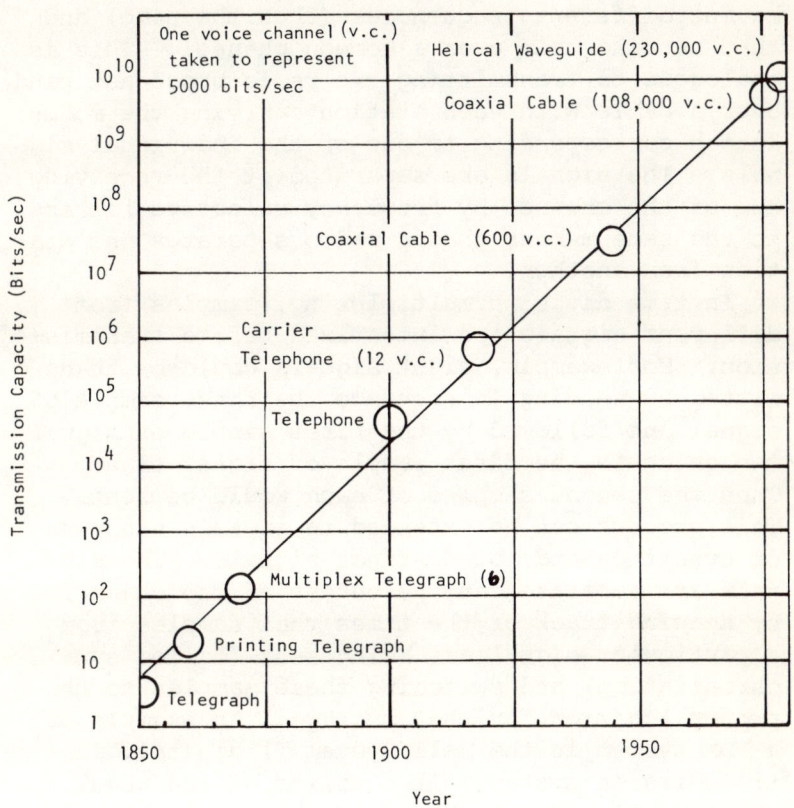

Figure 2-12   Development of telecommunications channel
              capacity (Adapted from J. Martin, *Telecom-
              munications and the Computer*, Prentice-
              Hall, Inc., Englewood Cliffs, N.J., 1969).

    In order to make good use of large capacity
channels, it is necessary to combine many separate
signals and send them together over a single channel. This is called multiplexing and is done in

many ways. Two of the most common methods are
frequency division multiplexing (FDM) and time
division multiplexing (TDM).

Frequency division multiplexing is accomplished
by modulating the individual signals onto carriers
having different frequencies (thus the name) and
then sending them over a common channel. This is
analogous to transmitting the radio broadcast band
over a cable with each station carrying the modulation corresponding to one of the individual signals. The signals are separated at the receiving
end of the channel by frequency selective filters
in the same manner that a radio separates one station from another.

In time division multiplexing, samples from
different signals are intermixed before transmission. For example, eight signals could be transmitted by sending in sequence the first sample of
signal one followed by the first sample of signal
two on up to the first sample of signal eight.
Then the second samples of each would be sent.
This process can be extended to include hundreds
or even thousands of distinct signals. The signals are separated at the output of the channel
by keeping track of the times that samples from
a particular signal are being sent (called syn-
chronization) and switching these samples to the
proper location for that message. An example of
a TDM system is the Bell System T1 digital data
transmission system. It consists of the regular
twisted pair of telephone wires but with closely
spaced repeaters. It is capable of handling 24
speech channels of 8 bits quantization. This
corresponds to 2 WN x 24 = 2 x 4000 x 8 x 24 = 1.5
million bits/sec. The next planned system, the T2
carrier, would operate at a bit rate of 6.3 M
bits/sec and could accommodate four T1 carriers or
one picture phone signal.

Typical Information Source Data Rates

The human mind and body are only able to accommodate inputs at relatively low rates, nothing like
the rates presently possible in telecommunication
channels. For instance, consider the information
rate of human reading. A fast reader might cover
as many as 1,000 words per minute. This corresponds to 83 letters per second or 415 bits per
second. Display screens available today for pre-
senting printed text operate at 2,400 bits per
second, more than five times reading speed. Computer printers, the devices that produce the hard
copy output, can generate 1,200 lines per minute
which is 20,000 bits per second. This is also the
speed at which IBM cards can be read.

A PCM voice channel with 4 kHz bandwidth and
8-bit quantization generates 64,000 bits per second. Note the great discrepancy of more than 150:1
between the required transmission rates for reading
and talking. If the word content of a message was
all that was important, it would be possible to improve communication efficiency greatly by using
teleprinting instead of telephoning. Actually, the
other aspects of speech communication such as voice
inflections, familiarity, personal awareness, etc.,
provide dimensions not readily obtainable in the
printed words or artificial generation of speech
at remote locations.

High fidelity sound reproduction represents a
substantial increase in data rate for the communication channel. For a bandwidth of 20,000 and a
quantization level of 10 bits to assure high quality reproduction, the transmission rate would be
2 x 10 x 20,000 = 400,000 bits/sec. The next step
beyond high fidelity sound reproduction is television. For commercial television with 525 lines
resolution the required bandwidth is 4 MHz and
with 10 bit quantization this corresponds to 2 x
10 x 4 = 80 megabits per second. This is equiva-

lent to about 1,200 voice channels, and this
tradeoff is what actually occurs in satellite and
intercity communication channels used to handle
both television signals and telephone signals.

Telecommunication Channels:
Present and Future

Telephone communication between cities makes use
of coaxial cables and microwave relays. A coaxial cable consists of a copper tube with a wire
running down its center. Usually there is insu-
lating material between the outer tube and the
inner conductor. This material keeps moisture out
of the tube and holds the center conductor in the
middle of the tube, which is necessary for proper
operation. The coaxial cable is used to carry
electromagnetic (radio) waves from one place to
another. The waves all stay inside the tube and
can be amplified or switched to other channels as
required. The primary limitations of the coaxial
cable are its bandwidth and the attenuation of
signals that necessitates amplification at short
intervals. A modern coaxial cable can handle
nearly 5,000 voice channels. Typically a number
of cables are bundled together and installed un-
derground as a communication trunk. Figure 2-13
shows 1973 coaxial cable installation and a cross-
sectional view of a trunk having a 108,000-voice
channel capacity.
   Microwave relays use signals radiated through
the atmosphere from antennas on one tower to antennas on another tower. By using very high carrier frequencies, about 4 to 6 GHz, it is possible
to obtain large bandwidths and simultaneously restrict the radiation to narrow paths so it will
not interfere with others using these same frequencies. The technology in these systems grew
from radar developments during World War II. The

Figure 2-13   Installation of a coaxial cable in the
              Mississippi River. Bell Labs Photo.

early systems built during the 1940's carried a
few hundred voice circuits; today a microwave relay operating simultaneously on two carrier frequencies can accommodate as many as 29,000 voice
channels. When current systems are further mod-
ernized, it is anticipated that the capacity will
be increased to 34,800.(5)   Figure 2-14 shows a
modern microwave relay tower used for intercity
telecommunications.
     The next step in long distance transmission was
the communication satellite which came into serv-

Figure 2-14   Microwave relay tower.   Western Electric
             News Features.

ice in the 1960's. Satellites are discussed in
considerable detail in Chapter 5 and will not be
considered here except to note for purposes of
comparison with microwave relay, that the domestic
communication satellites scheduled for operation
in 1976 will have 14,400 voice channels each and
will be accessible from several different ground
stations.
    The next major developments in surface trans-
mission of data may be the circular waveguide and
the optical fiber. The waveguide is essentially
a hollow pipe built to precise tolerances that
guides radio waves from one point to another with
very low attenuation. Signals can be transmitted
via waveguide for 32 km (20 miles) before they
must be amplified whereas for signals in coaxial
cables there must be amplifiers every 3.2 km (2
miles). Because they operate with very high car-

rier frequencies, (40 to 110 GHz) it is easy to
obtain very large bandwidths. Installation of the
first experimental system was begun in 1975 over
a 13.7 km (8.5 mile) segment in northern New Jer-
sey. This system will have an initial capacity
of 230,000 voice channels, a capacity which will
ultimately be expanded to 460,000 voice channels.
Figure 2-15 shows a segment of waveguide being
installed.

Figure 2-15   Section of circular waveguide being
              installed.   Bell Labs Photo.

Thus it is clear that the achievement of increased bandwidth is closely tied to higher carrier frequencies. The ultimate step in this direction appears to be the use of light waves as carriers of communication signals. The frequency of electromagnetic radiation corresponding to visible light is thousands of times greater than that of the highest frequency radio waves presently used for communication, and the available bandwidths are correspondingly larger.

Two developments in recent years have made optical communications feasible. The first is the laser, a powerful generator of pure and high power light waves. The second is the optical waveguide, a specially constructed glass fiber smaller in diameter than a human hair, that can conduct light from one point to another with low attenuation. Ultimately, these developments may lead to systems having Gigabit per second capabilities in a single fiber. A bundle of several hundred such fibers would only be the size of a pencil, but would provide a communication channel far better than any existing today. Very likely fibers will find their way into the home and office in the not too distant future as a broadband communication channel to serve a wide variety of needs. Figure 2-16 shows a bundle of optical fibers.

Solid State Electronics

While the evolution in transmission capacity of communication channels has been occurring, enormous and revolutionary changes have been occurring in the field of electronic devices and components. Development of what is generally called solid state technology brought about these changes. Extensive research on electronic applications of crystals as detectors of electromagnetic radiation was carried out during World War II. This work continued after the war and led to development, in

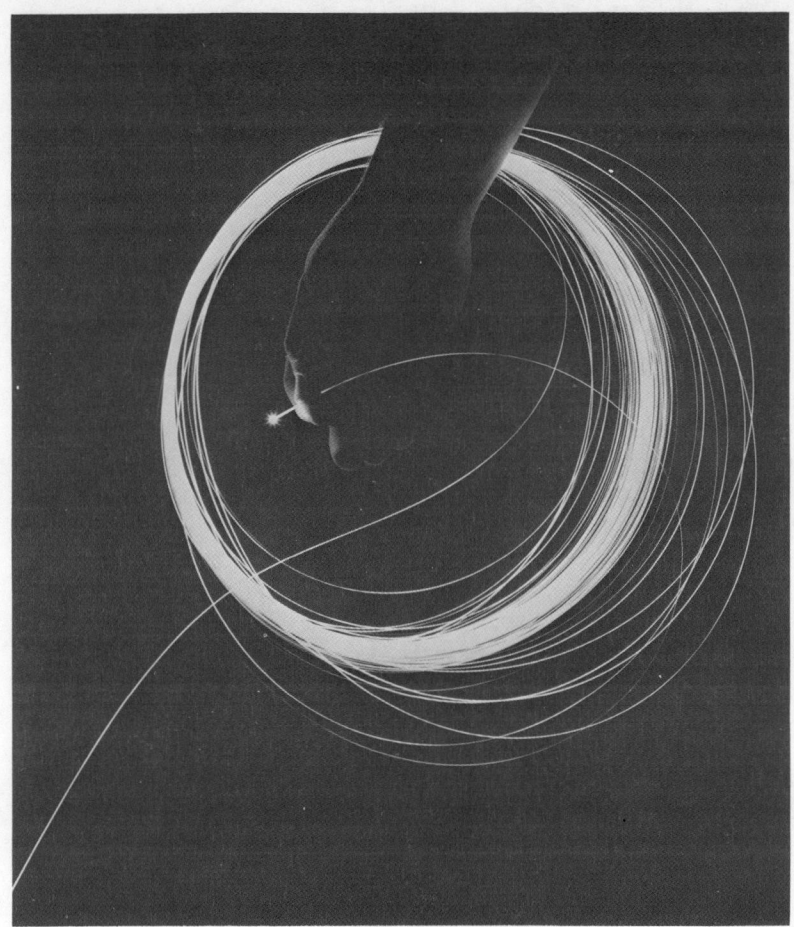

Figure 2-16   Optical transmission line.   Bell Labs
             Photo.

1947, of an entirely new kind of electronic amplifier, the transistor. This small device did not
require the electrical heating element necessary
to all vacuum tubes. From its inception this new
device was destined to have a major impact on the
entire electronics industry.

    Initially, the transistor could not compete in
any significant way with the vacuum tube. However,

by the middle 1950's sufficient advances had been
made in solid state technology so that transistors
could be produced that were more than competitive
with vacuum tubes. This was particularly true for
military electronic applications where small size,
ruggedness, and reliability were of utmost importance. The rapid development of solid state technology during the 1950's was due in large measure
to research and development efforts sponsored by
the Department of Defense.

Because of the inherently small size of transistors, producing miniaturized electronic components and subassemblies was emphasized. These
developments progressed until "integrated circuits," transistors and their associated components
and interconnections, were simultaneously produced
in a small package as a complete functional unit.
Instead of a transistor, one could purchase a complete amplifier or other component. Frequently,
the cost of these complete units was many times
less than that of a single transistor a few years
earlier. "Throw-away" portable radios became a
reality, radios, which could be purchased for less
than ten dollars and discarded if they were damaged or failed for some other reason. The cost of
repairing them exceeded their replacement cost.

The transistor and its successor, the integrated
circuit, found instantaneous acceptance in the
field of electronic computers. It is safe to say
that without solid state technology the computer
would not be the pervasive influence in our society
that it is today.

Another step in subminiaturization of electronic
components occurred with development of the capability of making large scale integrated circuits
(LSI). With this technique literally thousands of
transistors and associated components would be
placed in postage-stamp size packages. This led
to such new consumer products as pocket calculators
and digital wrist-watches. Figure 2-17 shows a

programmable pocket calculator that can remember
up to 100 instructions given it by the operator.

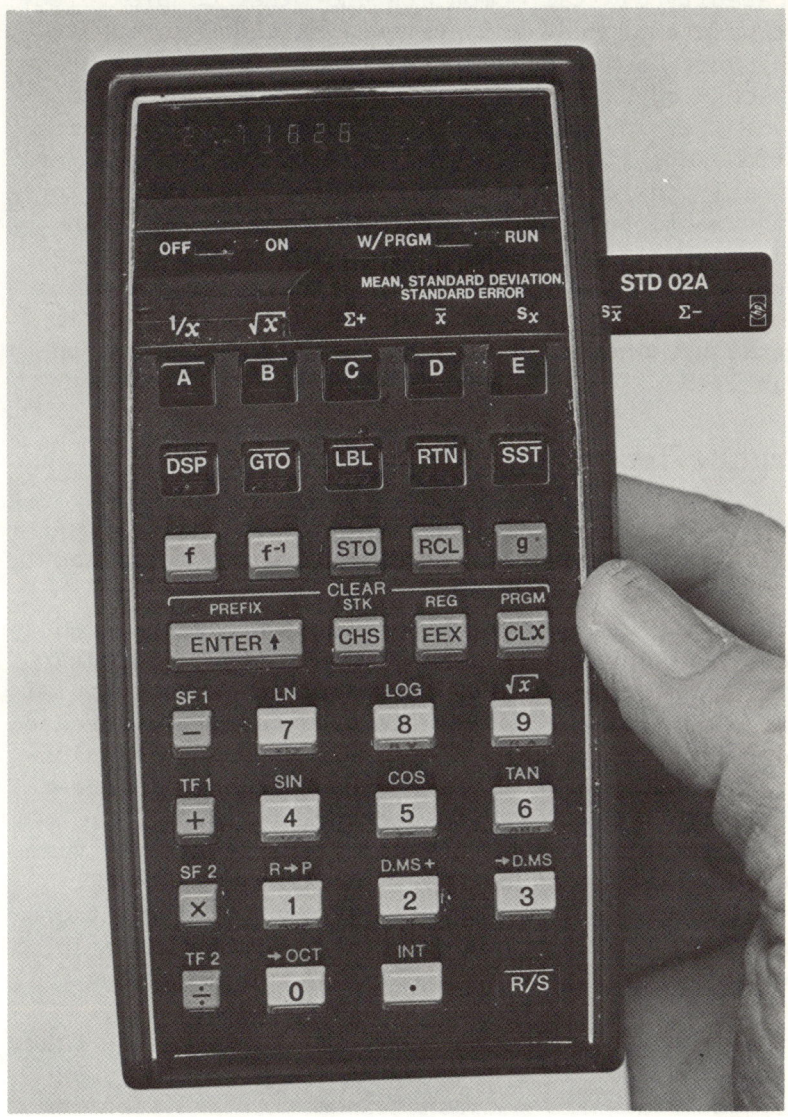

Figure 2-17   Programmable pocket calculator

This unit contains the equivalent of 70,000 transistors. More important, however, LSI technology has greatly reduced the size and cost of large capacity digital computers and other complex electronic equipment and systems. Without integrated circuits and the wide variety of other solid state technology that has developed in the past two decades, it is unlikely that we would have communication satellites, electronic switching systems, or nearly the communication capacity that exists in the world today. Through this technology, telecommunications services are being made available throughout the world at costs that put them within reach of large segments of the world's population that have never had anything like these services until now.

Future Technological Developments

The advent of inexpensive telecommunications devices and systems based on solid state technology foretells developments of great potential within the next few decades. The two most obvious trends are the continually expanding use of the computer throughout government, business, and society in general, and the greatly reduced cost and increased availability of long distance telephone communication. Other directions of technological innovation are equally as interesting.

The technological potential exists for the wired city to become a reality. The transmission of telecommunications signals between fixed locations does not require use of radiated electromagnetic energy as in radio and television broadcasting. Such signals can be carried more efficiently, though not now less expensively, by cables or other transmission lines than by radiated signal. Carriage of signals between transmitter and receiver would free major portions of the radio spectrum for other use. Potentially the number of channels

and the bandwidths available could far exceed anything possible by means of directly radiated signals. Also, talk-back facilities could be readily
provided for a variety of purposes ranging from
computer-aided instruction to voting. A 1975 study
described possible services that could be provided
in the home by a wideband telecommunications terminal.(6) Estimates of the probable interest and
usage of such services and when they might be introduced to society were obtained from experts
using the Delphi technique of technology forecast-
ing. See Tables 2-6 and 2-7 for a list of potential services and forecasts. Although assessing
the accuracy of such predictions is difficult,
appreciating the enormous potential for such technological development is easy.

Other developments ranging from pocket telephones to automatic automobile driving and computer diagnosis of ailments are easily imagined.
But predictions can go awry, because many forces
other than technological feasibility can strongly
influence the course of development. These forces
are economic, social, and political, and their impact on technological development will be the concern of much of the next three chapters. Only
after examining these forces in detail will a return to consideration of the continuing evolution
of telecommunications technology and its impact
on society be possible.

Table 2-6   Brief descriptions of potential home
            information services (6)

1.   CASHLESS-SOCIETY TRANSACTIONS. Recording of any financial transactions
     with a hard copy output to buyer and seller, a permanent record and up-
     dating of balance in computer memory.

2.   DEDICATED NEWSPAPER. A set of pages with printed and graphic informa-
     tion, possibly including photographs, the organization of which has
     been predetermined by the user to suit his preferences.

3.   COMPUTER-AIDED SCHOOL INSTRUCTION. At the very minimum, the computer
     determines the day's assignment for each pupil and, at the end of the
     day, receives the day's progress report. At its most complex, such a
     service would use a real-time, interactive video color display with
     voice input and output and an appropriate program suited to each
     pupil's progress and temperament.

4.   SHOPPING TRANSACTIONS (STORE CATALOGS). Interactive programs, perhaps
     video-assisted, which describe or show goods at request of the buyer,
     advise him of the price, location, delivery time, etc.

5.   PERSON-TO-PERSON (PAID WORK AT HOME). Switched video and facsimile
     service substituting for normal day's contacts of a middle class mana-
     gerial personnel where daily contacts are of mostly routine nature.
     May also apply to contacts with the public of the receptionist, doctor,
     or his assistant.

6.   PLAYS AND MOVIES FROM A VIDEO LIBRARY. Selection of all plays and
     movies. Color and good sound are required.

7.   COMPUTER TUTOR. From a library of self-help programs available, a com-
     puter, in an interactive mode, will coach the pupil (typically adult)
     in the chosen subject.

8.   MESSAGE RECORDING. Probably of currently available type, but may in-
     clude video memory (a patient showing doctor the rash he has developed).

9.   SECRETARIAL ASSISTANCE. Written or dictated letters can be typed by a
     remotely situated secretary.

10. HOUSEHOLD MAIL AND MESSAGES. Letters and notes transmitted directly to
     or from the house by means of home facsimile machines.

11. MASS MAIL AND DIRECT ADVERTISING MAIL. Higher output, larger-sized
     pages, color output may be necessary to attract the attention of the
     recipient--otherwise similar to item 10, above.

12. ANSWERING SERVICES. Stored incoming messages or notes whom to call--
     possibly computer logic recognizing emergency situation and diverting
     the call.

13. GROCERY PRICE LIST, INFORMATION, AND ORDERING. Grocery price list is
     used as an example of up-to-the-minute, updated information about per-
     ishable foodstuffs. Video color display may be needed to examine
     selected merchandise. Ordering follows.

14. ACCESS TO COMPANY FILES. Information in files is coded for security:
     regularly updated files are available with cross-references indicating
     the code where more detailed information is stored. Synthesis also may
     be available.

15. PARTS AND TICKET RESERVATION. As provided by travel agencies now but more comprehensive and faster. Cheapest rates, information regarding the differences between carriers with respect to service, menus, etc. may be available.

16. PAST AND FORTHCOMING EVENTS. Events, dates of events, and their brief description, short previews of future theater plays, and recordings of past events.

17. CORRESPONDENCE SCHOOL. Taped or live high school, university, and vocational courses available on request with an option to either adult or graduate. Course on TV, paper support on facsimile.

18. DAILY CALENDAR AND REMINDER ABOUT APPOINTMENTS. Prerecorded special appointments and regularly occurring appointments stored as a programmed reminder.

19. COMPUTER-ASSISTED MEETINGS. The computer participates as a partner in a meeting, answering questions of fact, deriving correlations, and extrapolating trends.

20. NEWSPAPER, ELECTRONIC, GENERAL. Daily newspaper, possibly printed during the night, available in time for breakfast. Special editions following major news breaks.

21. ADULT EVENING COURSES ON TV. Noninteractive, broadcast mode, live courses on TV -- wider choice of subjects than at present.

22. BANKING SERVICES. Money orders, transfers, advice.

23. LEGAL INFORMATION. Directory of lawyers, computerized legal counseling giving precedents, rulings in similar cases, describing jurisdiction of various courts and changes of successful suits in a particular area of litigation.

24. SPECIAL SALES INFORMATION. Any sales within the distance specified by the user and for items specified by him will be "flashed" onto the home display unit.

25. CONSUMERS' ADVISORY SERVICE. Equivalent of Consumer Reports, giving best buy, products rated "acceptable," etc.

26. WEATHER BUREAU. Country-wide, regional forecasts or special forecasts (farmers, fishermen), hurricane and tornado warnings similar to current special forecast services.

27. BUS, TRAIN, AND AIR SCHEDULING. Centrally available information with one number to call.

28. RESTAURANTS. Following a query for a type of restaurant (Japanese, for instance), reservations, menu, prices as shown. Displays of dishes, location of tables, may be included.

29. LIBRARY ACCESS. After an interactive "browsing" with a "librarian computer" and a quotation for the cost of hard copy facsimile or a show-scan video transmission, a book or a magazine is transmitted to the home.

30. INDEX, ALL SERVICES SERVED BY THE HOME TERMINAL. Includes prices or charges of the above, or available communications services.

Table 2-7   Summary of Median forecasts (6)

Service	Average $ Value of One Conversation			Duration of Single Trans- action (min.)	Data Trans- mission Connect Time (%)	Avg. No. of Trans- actions/ Mo./Home
	Low	Middle	High			
1. Cashless society transactions	$0.10	$0.16	$ 0.40	0.75	20%	40
2. Dedicated newspaper	0.10	0.20	0.50	10.00	95	30
3. Computer-aided school instruction	0.50	1.50	3.50	30.00	20	20
4. Shopping transactions (store catalogs)	0.20	0.50	1.00	6.00	40	10
5. Person-to-person (paid work at home)	0.50	1.50	5.00	20.00	50	60
6. Plays and movies from video library	0.60	2.00	5.00	90.00	100	10
7. Computor tutor	1.00	2.00	5.00	30.00	20	10
8. Message recording	0.20	0.35	1.00	3.00	75	7
9. Secretarial assistance	0.35	1.00	3.00	10.00	60	10
10. Household mail and messages	0.10	0.20	0.50	2.00	90	25
11. Mass mail and direct advertising mail	0.10	0.17	0.50	3.00	90	25
12. Answering services	0.10	0.20	0.50	2.00	80	20
13. Grocery price list, information, and ordering	0.20	0.35	0.50	5.00	80	15
14. Access to company files	0.30	0.60	2.00	5.00	65	10
15. Fares and ticket reservation	0.20	0.35	0.75	5.00	50	5
16. Past and forthcoming events	0.10	0.20	0.50	4.00	80	10
17. Correspondence school	1.00	2.00	5.00	40.00	85	10
18. Daily calendar and reminder of appointments	0.10	0.20	0.50	1.00	80	25
19. Computer-assisted meetings	1.00	2.00	5.00	30.00	40	5
20. Newspaper, electronic, general	0.20	0.50	0.75	10.00	95	30
21. Adult evening courses on television	0.60	1.00	5.00	50.00	95	10
22. Banking services	0.10	0.25	0.50	2.00	60	20
23. Legal information	1.00	5.00	15.00	10.00	75	5
24. Special sales information	0.20	0.50	1.00	4.00	70	10
25. Consumers' advisory service	0.25	0.50	1.00	5.00	70	10
26. Weather bureau	0.10	0.20	0.50	1.00	90	20
27. Bus, train, and air scheduling	0.10	0.20	0.50	1.25	80	5
28. Restaurants	0.10	0.20	0.50	3.00	80	5
29. Library access	0.50	1.00	2.00	10.00	90	5
30. Index, all services	0.10	0.20	0.50	3.00	80	10

% of Service Home Subscriber Expected to Pay	Most Likely Year of Introduction			% Penetration of All U.S. Households	Median Transmit Time (min.)	Average Value of Service, $/Subscribing Household/Mo.		Value of Service After 5 Years, $/Average U.S. Household/Mo. (At Penetration Rates Shown)
	Early	Middle	Late			Median	High	
25%	1975	1980	1990	20%	0.19	$ 6.4	$ 12.50	$ 1.00
75	1980	1983	1990	10	9.00	6.0	15.00	0.54
50	1975	1982	1987	10	10.00	40.0	100.00	3.75
25	1977	1985	1990	10	2.50	3.0	5.00	0.38
5	1980	1985	1990	5	6.50	75.0	250.00	3.20
80	1975	1980	1985	10	90.00	20.0	50.00	2.00
80	1975	1980	1990	5	6.00	20.0	50.00	1.50
90	1975	1980	1985	10	2.50	2.5	6.25	0.20
100	1975	1980	1985	5	6.00	10.0	25.00	0.25
75	1980	1985	1990	10	1.80	5.0	12.00	0.60
0	1980	1990	1995	10	2.55	4.0	15.00	0.50
100	1975	1980	1985	10	1.60	5.0	11.25	0.30
50	1975	1980	1990	10	4.25	5.0	7.50	0.26
1	1980	1985	1990	2	3.50	15.0	50.00	0.30
40	1975	1980	1985	5	2.50	1.0	2.50	0.05
50	1975	1982	1990	5	2.70	2.0	5.00	0.11
75	1978	1984	1990	5	30.00	20.0	50.00	0.75
100	1980	1983	1985	5	0.95	4.0	10.00	0.20
40	1975	1980	1985	5	6.00	15.0	91.00	0.75
75	1980	1985	1990	5	9.00	15.0	22.50	0.75
80	1975	1980	1985	10	45.00	10.0	25.00	0.88
60	1975	1980	1985	10	0.85	4.0	10.00	0.38
100	1980	1985	1990	3	7.50	6.0	25.00	0.25
80	1975	1982	1990	5	2.70	5.0	10.00	0.25
100	1975	1980	1985	5	3.50	7.5	10.00	0.40
100	1975	1980	1980	5	0.90	2.0	5.00	0.12
80	1975	1977	1980	5	1.00	0.5	1.00	0.06
60	1975	1980	1985	5	2.50	1.0	2.50	0.05
100	1980	1985	1990	5	9.00	5.0	10.00	0.25
50	1975	1980	1985	5	2.50	3.0	5.00	0.10

$20.12

Chapter 3

Politics, Economics, and Telecommunications Technology

One major set of factors that strongly influences growth or retardation of technological development is the political and economic systems within which such development occurs. It is the political system that determines which interests receive encouragement and which are deterred from various actions. The policies which are made by portions of the government directly affect which technologies will be used for particular services, which technological developments will be encouraged, and which will not be fostered.

Economic considerations also provide direction for the development and use of telecommunication technology. Operating in a free market would provide the telecommunications industry with many guides about what services would be acceptable to the largest markets, at what prices. Such directions would certainly influence the industry's development of new technologies, utilization of existing ones, and provision of services. However, the marketplace does not always operate freely or perfectly, and the role of government regulation of the market and the industry has substantially altered these kinds of parameters for the operation of the industry. While the book will discuss these economic considerations as part of the examples covered in Chapters 4 and 5, it is useful to explore some of the general economic considera-

tions in this chapter, before more detailed discussions.

The chapter will present a general model for
political decision making which will provide a
basis for the analysis of particular cases which
follow in subsequent chapters. No social science
model can be precise, and many cannot suggest
casual relationships. The model developed here
is presented only as one way in which technopolitical decisions (political decisions relating
to technology) can be viewed. It should provide
a perspective from which the development of telecommunications technology can be considered.

Politics

Politics has been defined in a variety of ways,
but it has one widely recognized definition: It
is the authoritative allocations of values for a
society.(1) This definition contains several important points about the political scene.

First, politics is an *allocating process* by
which a variety of scarce values are allocated or
distributed among members of contending factions
of society. There are other allocative processes,
such as the economic marketplace. However, some
items cannot be allocated by selling them to the
highest bidder. Either because a price cannot
be fixed for some items or because people feel a
nonmarket procedure would result in a more equitable distribution of a good, the political process
is given the responsibility for many allocations
in a society.

Second, the political process results in an
authoritative allocation, one that the population
will respect and abide by. The authoritative-
ness of a governmental allocation may depend on
agreement with the allocation by members of society, or on the physical force which the government can bring to bear on recalcitrants to accept

the decision. Generally, however, the authori-
tativeness of a decision rests on a general level
of support for the government, its institutions,
and the current officials which go beyond the
particulars of the immediate decision. Thus,
most people accept most government decisions,
even if some people do not agree with the decisions
or benefit as a result of them.

Lastly, the allocation process operates on
values in the society. This means that whatever
the society places value on-- money, prestige, or
sea shells--can be allocated by governmental
processes. The entire allocation process can be
considered as one of allocating resources, as
long as the term "resource" is viewed in general
terms. In the telecommunications setting, the
kinds of resources which government allocates
include radio spectrum for broadcasting and other
uses, money in the form of government research
and development contracts, and permissible rates
which companies can charge for providing services
to the general public. These resources are not
the only ones the government allocates, but they
suggest the variety of items allocated. In general, the participants in the allocation process
view the resource involved as important and frequently valuable. The kinds of resources allocated will be illustrated throughout this book,
and will demonstrate the dimensions of governmental operation.

Political Arenas

A government is formed of institutions which are
designed to make the allocations discussed above.
The institutions are outlined by a constitution,
and the major importance of a constitution is
the procedural framework which it provides for
allocating resources. In the United States,
various governmental institutions make alloca-

tions involving telecommunications. The procedures which these institutions use depend on the
constitution or statutory law, and they vary
depending on the issue (or resource) and the circumstances (or politics) of the allocation.

The institutions and arenas discussed here
should be viewed as alternatives among which most
interested parties can choose. That is, each
arena is charged with making various kinds of
policy determination, and an interest which is
unsuccessful or noninfluential in one arena can
seek redress in another. As the book will outline, each institution cannot do the same thing
or make the same decision, but each can provide
some opportunity for allocation. Many of the
actors involved in telecommunications policy-
making value the alternative arenas available
and use them when they feel they can obtain a
favorable decision from any one of them. In
addition, the kinds of policies vary from arena
to arena. Thus, a formulation by Congress will
take a different form and have different consequences than a court decision or a decision by the
FCC.

Regulatory Agencies

Agencies of the government which are not mentioned
in the Constitution perform the primary allocation
functions relating to the field of telecommunications. The Federal Communications Commission, the
Interdepartmental Radio Advisory Committee, and
the Office of Telecommunications Policy all play
different, but important, roles in the political
process, yet none of these have constitutional
status.(2,3,4) They have been created by act of
Congress or by executive order, and they have
legal or advisory authority which place them in
central positions with regard to allocation.

The major regulatory arena in telecommunications is the FCC which has been given statutory

authority to regulate interstate wire and radio
broadcast operations by private companies in this
country.(5)   Since 1934, when it was created, the
Commission has issued general rules governing the
conduct of business by common carriers, such as
interstate telephone companies, and by broadcasters, such as AM and FM radio and television
stations.(6)   In addition, the FCC regulates the
use of the spectrum by individuals or companies
who seek enjoyment or profit through communications.   These include mobile radio users, such as
airlines and maritime companies, and citizens
operating shortwave broadcast stations.   The primary focus of this book is on the major commercial
users of radio spectrum, such as broadcasters, and
on the wire common carriers.   Additional details
about FCC authority are contained in the Appendix
where portions of the Communications Act of 1934
are presented.

The major segments of the industry which the
FCC regulates involve large companies with vast
resources of money and technology. The regulation
of these directly influences the development and
utilization of technology. Many observers have
charged, however, that the FCC has been "captured"
and is controlled by the clientele which it is
supposed to regulate so that its allocative decisions reflect those interests rather than the interests of the general public.(7,8)   The statutory
guidance for the agency is that it is to regulate
in "the public...interest."(9)   However, this vague
phrase has no precise definition and gives the
agency wide discretion to make whatever decisions
it wishes.   Since there has never been a precise
definition of the public interest, the usual approach is to define the public interest in terms
of whatever considerations and interests one
wishes.   Under this guidance, either the public
interest is what a particular company wants if it
is the only party involved in the agency proceed-

ing, or it can be a mixture of all the interests or forces involved in the allocation process for a particular policy issue, including competitors, citizens groups, and others.

Generally, the agency has tried to balance various competing interests on an issue before reaching a decision. This may be nearly a super-human task since some of these enterprises may be diametrically opposed to others or the variety of them may be great. It is interesting to note that even if the FCC has repeatedly sought to balance these interests and not injure any one of them greatly, the agency has most frequently not in-jured the vested interests of the clientele it regulates. Except in recent years (which may be the exception to the rule) the agency has never made decisions which consciously conflict with the interests of established common carriers or broad-casters. Despite isolated decisions which result in some disadvantage to a particular member of the regulated clientele, the agency's general policies tend to reflect this effort at compromising and protecting the clientele.(4)

Whether the agency should balance interests differently than it has, is an engaging and impor-tant policy question. The answer depends largely on the policy the individual observer prefers. It is important to observe that the agency is a cen-tral arena for the allocation of resources in the telecommunications industry. It is the arena where most of the interests focus most of their efforts. Although lobbying is not a precise term, these groups do devote much of their lobbying ef-forts to the agency and its personnel. There is a nexus between industry jobs and agency jobs, and the transfer of people between the agency and the industry tells an important part of the regulatory story because high-level personnel do transfer easily and frequently from one to the other.(7)

Furthermore, the regulated interests do petition the agency for various policy decisions at
various times. This direct petitioning is not
lobbying but rather direct requests of the agency
to exercise its allocation functions. The agency,
using various procedures, considers these requests
and usually acts upon them. Despite charges of
delay and procrastination, the FCC's decisions
usually respond in some fashion to the demands
made upon it. Often these responses acquiesce in
a vested interest's requests or deny a challenger's
petition.

The Legislature

The legislative body, the U.S. Congress, is involved in telecommunications allocations in various ways. Obviously the enactment of a statute
requires a good deal of Congressional action and
effort by many people. However, in less direct
ways, Congress fosters allocations, controls them,
or makes suggestions. Through a statute, Congress
can specify which industries will be regulated and
by what means. In rare instances the legislature
can make an allocation itself, outlining which
procedures and which interests will receive what
resources.

Congressional procedures are complex and are
often cumbersome and slow. The complexity of most
modern legislation requires that the Congress
specialize by subject matter. Thus, various committees perform the primary functions of screening
proposed legislation on topics such as telecommunications.(10,11) These committees, composed of
selected members of Congress, may review agency
decisions, may hold hearings on proposed legislation, and may repeatedly review the budget proposals of the agencies. Thus, through such legislative oversight, Congress exerts pressure on
agency allocation processes.(12) Although Congress

monitors what agencies do through budget hearings
and annual reviews and is thus in a position to
make necessary adjustments, it does not meddle in
the allocation process in most circumstances. The
agency may respond closely to the suggestions Congressional committee members make, since the life
blood of the agency, its budget, must be approved
by these committee members.

On rare occasions, Congress may become so concerned about a particular policy that it will
proceed beyond the oversight function. Thus, the
appropriate committees might seriously consider
proposed legislation with an eye toward changing
the situation Congress is concerned about. Some
examples of this in the telecommunication field
will be discussed in later chapters. However,
remember that enacting legislation is the extreme
form of Congressional activity, and it is actually
the rare case. The legislative process is quite
slow and is filled with many obstacles which substantially reduce the likelihood of enacting legislation. Only if a proposal is strongly supported
by the White House, a large group of Congressmen,
the private lobbies, or some combination of these
is a proposal likely to reach any stage beyond being introduced and referred to committee. With
various coalitions of support, however, a bill may
be enacted. This is a relatively clear legislative
pronouncement of purpose and policy preference.
Most legislation tends to be general and sometimes
is unclear, so that interpretation of policy must
occur by the executive, by the agencies applying
a statute, or by the courts which may be called
upon to interpret the meaning of the statute.
However, any Congressional statement of policy in
the form of a law is more formal and precise than
many other political influences which operate on
policy matters.

The political forces which operate on Congress
are much more numerous than those the FCC feels.

The same interests which lobby before the FCC also
lobby before Congress. However, in the legislative arena they are much less the center of the
process since their interests do not necessarily
have great weight with Congress, and many other
interests compete for Congressional attention.
Various telecommunications lobbies do surface and
do become very active when legislation is being
considered which will impact on their interests.
Interests have attempted to achieve legislative
action on a policy when the FCC has decided against
the interest, and the FCC has lobbied for some
legislation at times.(4) The congressional-agency
relationship is largely in terms of the oversight
function mentioned above. However, there are clear
cases in which the relationship is the more traditional legislative, lawmaking function. Some of
these efforts are illustrated in the following
chapters.

The Executive Branch

The role of the chief executive in telecommunications politics is not unlike the role played in
other policy areas. The President, as head of the
Executive Branch, relies on staffs to assist him
through policy suggestions, proposed legislation,
analysis, and administration regarding the subject
matter area.(13) The President cannot possibly devote any major portion of his time and decision
making to a particular policy area, and only when
a problem arises, a large and vociferous interest
surfaces, or some apparent need for action occurs,
does the President become directly involved in
such policy making.

Beyond the President, however, there is a bu-
reaucracy within the Executive Branch which spe-
cializes on the telecommunications industry and
devotes full time to considering issues arising
in the area. The structure of the staff varies
with each President, however. The Office of Tele-

communications Policy (OTP), and the Office of
Emergency Planning have been recently structured
to serve the President as advisers on the subject
and as analysts of policy proposals.(14) These
arms of the President can be very effective in
reaching policy agreements with industrial concerns, proposing legislation, and developing budget proposals for the implementation of such policies.(15)

In traditional terms, the Executive Branch is
intended to administer policies, and in these terms
the appropriate cabinet members may use staffs of
expert assistants to work on communications problems which relate to each department's operations.
Thus, the Department of Defense requires a large
group of communications advisers to deal with
problems of national security and military command
relating to communications systems. The Department of State must deal with various problems of
communications relating to international relations
and our conduct of foreign policy. The Departments of Commerce, Transportation, Housing and Ur-
ban Development, and Justice all have groups which
consider and work on problems of communications
which relate to the missions of these departments.

One of the major communications functions of
the Executive Branch is in the role of policy formulation. This involves the President in the formulation and suggestion of major policies in all
areas of government concern including communications. The executive has become the major initia-
tor of legislation, and, as a result, the further-
ance of a telecommunications policy through enactment of a law is likely to require White House
approval if not active support. Through the
executive functions of budget control and planning,
the White House has become central to the initia-
tion, as well as the administration, of policies.

The result of this is that the White House is
subject to various kinds of political lobbying.

While the lobbying efforts relating to the White
House are different than legislative lobbying tactics, the same advocates and participants may be
involved. The access patterns for interests differ and that explains much of the difference with
regard to White House lobbying. An interest's access to the White House is much more closely tied
to the preference of the President (the individual),
than in Congress, where nearly any interest can
probably find one or more sympathetic legislators
out of 535. Thus, the arena chosen for lobbying
depends on where the interest has the best access,
and the White House is one of a number of arenas
which give opportunity to various telecommunications interests on occasion.

Beyond the relatively formal functions of initiating and administering legislation, the Executive Branch provides opportunities for political
compromise and settlement among various contending
interests. The President and his staff perform
the role of mediator and broker.(15) Just as Congress plays a major role in controlling agency
policy through oversight, so the executive can
formulate and get agreements among interests by
talking with them and getting them to talk to each
other. While the President may be able to influence the decisions of administrative agencies, the
regulatory agencies are supposed to be independent
of partisan political influence. The President,
once he has appointed the members of the Commission, may have little access to the agency's decisions. However, in its own right, the power of
the Executive Branch places it in a central position to advocate positions, suggest compromises,
and, in some cases, dictate results, when there
are major policy differences among factions or
groups of interests.(14) As some of the following
cases will illustrate, this informal brokerage may
be the most important role performed by the executive in terms of policy formulation and operation.

The Judiciary

While in formal, constitutional theory the judiciary is not intended to make policy, the courts
in this country do perform policy making functions
by interpreting statutes and by deciding other
cases involving the law.(16,17) Given a factual
situation, the court can be presented with significant policy questions (phrased in terms of legal
questions) which impact directly on developments
in the telecommunication field. Some examples of
this include the Supreme Court's decision giving
the FCC jurisdiction to regulate cable TV as an
ancillary to the wire and broadcast regulation
given the commission by the statute. Others are
repeated court decisions, that, under the then
existing copyright laws, cable TV companies are
not required to pay copyright fees to the holders
of the copyright privileges for television programs carried by the cable.(18,19)

Although the courts are not directly subject to
political lobbying as is Congress, the President,
and the FCC, the judiciary can be a selected political arena in which an interest will seek to vindi-
cate its position or achieve a policy outcome.(20)
An interest can litigate a case, in order to gain
an advantage over a competitor, rather than lobby
Congress for the enactment of a statute. While the
primary examples of judicial lobbying are found in
civil rights litigation involving minority interests which could not get a satisfactory hearing
before Congress or the other branches, communications interests which lose in the agency or Congress, can and do "appeal" to the court for reversal of the adverse decision. The examples discussed
later in this book illustrate some of these efforts, and in a few instances their success.

While most people may not consider the court as
a viable arena for policy making, the only requirements for getting judicial treatment of an issue

is to develop a "case" which involves a legal
question. Thus, any set of facts which can present a legal dispute between two parties can be
taken to court for a resolution. Lawyers are
trained in framing legal issues, and nearly any
major policy question in communications policy (in
any subject matter for that matter) can probably
produce at least one legal question. Whether the
party will want to litigate the question depends
on the amount of time, money, and interest in the
question and on the likelihood of winning in the
court. However, it is quite possible to construct
such a case legitimately in order to seek a favorable policy result which could not be obtained in
the other political arenas.

Political Actors

From this discussion, we can see that many different interests and policy positions appear in
the telecommunications area. The actors generally
tend to have a stake in the outcome of governmental decisions. They seek favorable decisions by
many different means, some legitimate and some
possibly illegitimate. They are not always consistent, they may conflict directly with one another, or they may overlap on some matters and form
strong, momentary political coalitions. This is
clearly the general proposition of politics--the
competition of various interests for the allocation
of scarce resources. However, the examples provided in this book will involve a relatively small
number of interests. The actors will become familiar to the reader, even if their positions are not
always the same or even always clear. It may not
be necessary to provide much more here than a listing which the reader can use like a score card.
However, it is important to describe, in general
and specific terms, the kinds of interests which

appear and the kinds of tactics they employ to
achieve their desired ends.

Generally, interests in this area do not contend
in the political arena for technological decisions.
That is, they do not advocate the choice of technologies which the government then settles on as
the appropriate means for processing certain communications. Such technology allocations may
occasionally arise as in the case of competing
systems of color TV in the early 1950's, but they
are infrequent. Largely, the interests compete in
the political arenas for economic advantages over
competitors. The kinds of issues which arise often
tend to be basic economic ones, and the tactics of
the participants tend to be direct, obvious, and
frequently strong because the stakes are significant for them, if not for all of society.

The kinds of interests which surface in this
area are not unlike those in other areas. The most
active and successful ones tend to be those which
have a clear position and policy preference and
which are directly concerned and affected by the
policy. Recently, more diffuse interest groups
have surfaced which appear to have sufficient
strength and political resources to compete in the
political arenas. Consumer groups and groups of
users or viewers of telecommunications services
are many in number, difficult to organize, and so
diverse in interest that it is unlikely they will
be able to pursue a consistent policy effort
through lobbying. If they could, however, they
would probably be a strong political force because
there are so many of them. Policy makers tend to
listen to large numbers of voters, because their
continued political careers depend on being re-
elected, so consumers could exert great pressures
on elected officials, even if not on appointed
agency members. Only in a few policy areas have
such diffuse groups voiced much interest, and that
has occurred only in recent years.

It should be apparent that the kinds of interests involved are large groups of diffuse member-
ship with only minimal or tangential and passing
interest in an issue. In contrast to these kinds
of orientations, a smaller group of companies is
involved in the communications industry and is directly and immediately affected by any policy made
in the area. These tend to be constantly active
or on watch and they appear at every opportunity
to deal with a policy consideration which concerns
them. These groups include the common carriers
and broadcasters who are themselves regulated by
the FCC and who compose the immediate, communications industry. In addition, peripheral interests
involve equipment manufacturers and specialized
users of portions of the spectrum who lobby constantly and sometimes successfully for policies
which favor their interests.

The Vested Interests

Under this category any company which had a stake
in the status quo would be included. Thus, the
vested interest in any particular situation depends on what the issues are and who has become
established in providing the service. Generally,
the most prominent vested interests include AT&T
which has all the long distance, and much of the
local voice telephone communication market in this
country. The company also is the principal international voice carrier, and a subsidiary of AT&T
(Western Electric) is the major developer and producer of telephonic equipment in the world. In
addition, the Bell Laboratories are a part of
AT&T and constitute the world's leading, private
research and development organization in the communications field. Other established common carriers include non-Bell telephone companies, Western Union which provides much of the record (nonvoice) communication service in the country, and

such international carriers as ITT and RCA. Among
broadcasters, the established interests include
the networks and the licensed radio and television
broadcasters.

While this group of interests may not seem to
be particularly large, remember that they form the
backbone of the communications industry in this
country and possess enormous resources for lobbying in political arenas. They also have a large
number of economic and technical interests which
give rise to many and frequent efforts at obtaining protection and favorable treatment from the
political actors. The broadcasting interests may
illustrate this point since they operate on the
basis of licenses issued by FCC assignment. Their
interests in these licenses are great since the
investment in broadcast equipment is great, and
their revenue from the sale of advertising time
suggests the amount of income they can expect from
the possession of the license. As a result of the
investment and the value of the license, broadcasters make major efforts to deal with threats or
challenges to the status quo. Some changes may
benefit them, such as current legislative proposals which would extend the license period from
three to five years. However, the broadcasters,
just as most other established interests, are
more on guard for threats to their position than
making efforts to improve it.

Interestingly, the established interest in any
specific instance will depend on who is defending
the status quo. As a company becomes protected or
has something which it seeks to protect, it becomes a vested interest in communications politics.
Thus, over time, if a new competitor is successful
in gaining entry to the market and succeeds in
making a profit, that company will become a political, vested interest which lobbies for the status
quo or continued protection of its position. Political and policy change will produce new estab-

lished interests seeking political protection or
continued advantage in the communications field.

The New Entrants

The other major set of identifiable interests is
comprised of the challengers who seek some changes
in the status quo. These are usually companies
which propose to offer a new service or provide
an existing service at competitive rates. These
are generally companies seeking to be permitted
to compete with the vested interests. Examples
are outlined in subsequent chapters, which indicate the reaction of established interests to
such threats. In recent years, the FCC seems to
be allowing some new entrants into arenas previously monopolized by the established carriers.
For many years, the FCC, as most other federal
regulatory agencies, made few decisions that dis-
turbed the established interests. Either because
of cooperation by the industry, or because the
commissioners felt that the established interest
could best perform their obligations without competition, the commission, until the mid 1960's,
made no decisions which did anything but reject
the various efforts of the challengers to enter
an established market.(21)
     Protection of the industry has been a major
political issue for many years, and there are
examples in which the courts, generally viewed as
very conservative bodies, reversed commission decisions protecting the industry and required the
agency to permit entry of new companies or permit
some action which the industry argued would injure
their position.(22) The political efforts of potential entrants depends on the amount of resources they have for lobbying and on the policy
climate in which they find themselves. Thus,
recently, new entrants have been receiving favor-

able decisions from the FCC itself, while in previous decades a challenger had to persevere beyond
the agency--into the courts, legislature, or exec-
utive--in order to succeed. The recent political
efforts of these interests have required less
money and time and have resulted in more successes
because of policy orientation changes in the
agency.

Beside new entrants, other challengers may be
groups of users. While we all use the telephones,
watch television, and listen to the radio, we
rarely would consider seeking some sort of allocation from the FCC. In recent years, however, the
users have begun to intervene frequently and with
some success in agency license renewal cases. Some
would claim that the commission's statutory objective of regulating in the "public interest" means
that the commission will, on its own, consider
such user interests and the users themselves have
little need to appear. However, with the rise in
consumerism and various public interest groups in
recent years, other observers would claim that the
agency ignores such interests unless they are presented in the proceeding in the same fashion as
the requests of the established interests or the
challengers.

There are other users besides the listeners and
watchers. Some companies use a large amount of
communication service on a continual basis. Thus,
a large company, with plants throughout the country, may lease a telephone line from AT&T which
is permanently dedicated to linking all the company's plants. These large users form a significant portion of the common carrier market, and as
large, continuing customers, the common carriers
usually seek to provide quality service to them.
There have been occasions when the commission's
decisions have been adverse to the carriers and
to large users of services, and in one particular

case the users, rather than the carrier, sought a
court decision which would protect the carrier's
position.(23) This same category of users can provide the basis for new entrants to seek licenses
to offer services. If the established carriers
are not providing a service or are doing so at
higher rates than a new entrant, the users who
would "switch" to the new entrant provide market
support for its claim of entry and provide the
obvious economic base upon which the challenger
seeks to make a profit. These users may be mobil-
ized for political action by the challenger to
support their petitions.

Another large group of "challengers" may more
properly be viewed as both vested interests and
challengers. These are equipment manufacturers
and developers who profit by the use of various
kinds of technology. These groups can benefit
greatly from various decisions which require the
offering of certain services, since this decision
creates the market for the devices produced by
these companies. The manufacturers of mobile
radio telephone equipment, for example, have been
at the forefront of efforts to get the FCC to allocate to mobile communications increasingly
larger portions of the radio spectrum. So have
producers of citizens band communications equipment in recent years. There are other examples
of these efforts by manufacturers which are illustrated by some of the cases examined in subsequent
chapters.

Note that these equipment manufacturers, just
as some established interests, have another set of
interests in the political arena. These companies
often make a good portion of their income from
sales of equipment to the government. The Department of Defense and other cabinet agencies *use*
telecommunications equipment and services and they
do not make it themselves. Government contracts
are a lucrative source of business, and, for these

manufacturers, such contracts often supply a major impetus for their research and development of new technological devices. For example, the use of geosynchronous communications satellites was made possible much earlier than expected because of government contracting with a manufacturer to develop such a working system (contrary to the technological development and position of the other private organizations developing communications satellites in the early 1960's).

One last group of challengers has appeared occasionally in communications policy issues. This group is composed of independent research organizations or nonprofit foundations which have some interest in the development of communications policies or investigate such a subject because there is some "public interest" concern which the organization seeks to represent. This group includes such organizations as the Ford Foundation, the Alfred P. Sloan Foundation, and the Committee for Economic Development. Some universities, such as Harvard, have developed programs which do research work for the government or other supporters which seek policy impact. Only isolated examples of these kinds of interests surface in the cases this book presents. However, they have played roles in several of the cases. Either because they are prestigious organizations, their policy recommendations are persuasive, or their position is close to the compromise position settled on by the policy maker, these groups can appear to have substantial influence. Rarely will such a group begin with a preconceived policy preference. Some of them shun policy choices, but present only their "analysis" of the issue. However, some do present policy recommendations at the conclusions of their investigation.

Public Officials

The last set of actors which is central to the
political allocation process is the persons in
public positions who have the obligations and duty
to make authoritative decisions. These are the
arena occupants and, in some sense, they are the
arenas, since institutions without people are
meaningless. Some of these officials are elected
and are "obligated" to represent a particular
group of constituents. Congressmen represent
either a state (in the case of U.S. senators) or
some sector of the state (U.S. representatives).
The President represents the entire nation. The
degree to which any of these constituencies can
be represented by a single person is questionable.
However, as a result there is a heterogeneous set
of representations made in the decision making
arenas, by the decision makers themselves as well
as by competing lobbying interests.

It may be impossible for an individual to determine the interests of a constituency, let alone
to represent them effectively in a political body.
Any geographic region which is represented in Congress has a multiplicity of interests, many conflicting and some overlapping. This makes it more
impossible for the elected official to present a
clear, accurate view of his constituent's interests.
Some representatives do not try to reflect their
constituents' views but simply use their own ideas
and values as guides for decisions. Other decision makers try to reflect constituents' views on
some issues, but on other matters use their own
judgment or follow what they think is the "public
interest."

Besides elected representatives, some decision
makers are appointed and thus are not responsible
to an electoral constituency. These appointees
(largely agency members and federal judges) are
not objective automotons but have preferences and
biases just as other people. However, they may be

less restricted in exercising those biases than are
the elected decision makers. While Federal Communications Commissioners and judges undoubtedly
feel that their roles require that they weigh and
balance a large variety of competing interests,
and they probably try to do this conscientiously,
they still may be less representative than elected
officials. To whom does a judge, appointed for
life, have to answer? No one specifically, even
if the judge feels largely constrained to do a
conscientious job of "judging" objectively.

Appointed by the President, FCC members serve
for seven-year terms, and the statute requires that
the members of the commission must be balanced to
include no more than four members of any one political party.(24) This last provision is designed to
protect against a completely partisan commission.
Many observers have charged that the agency's constituency may not elect them but they are likely
to be employed by the constituency (the communications industry) after their seven-year term ex-
pires.(7) There has been a good deal of interchange
of personnel between the Commission and the industry, and the decisions of the agency *may* generally reflect a pro-industry bias. However, these
kinds of considerations are probably not too important if the overall picture is considered.

The variety of officials who can and do make
decisions concerning communications policies is
great. There may be wisdom in this because it permits a wide set of interests to have some "representation" in the decision-making process. The resulting balancing of interests and various checks
on decision makers do not result in policies which
are likely to injure one particular segment of society repeatedly. Since competing interests have a
variety of arenas in which to seek resolutions favorable to them, and since many interests have
representatives among the policy makers, it may
appear that the political process is designed to
produce policies which balance competing interests.

International Politics

While most of the discussion presented in this
book involves domestic telecommunications policy
and domestic politics, international politics and
considerations are also of great importance. In
fact, international cooperation and diplomacy are
directly involved with many of the same questions
which will be discussed. International forces
arise because much communication itself is international in character. This requires arrangements
for such transmissions. In addition, the radio
spectrum and problems of its allocation for various users and uses involve international issues
since radio interference does not stop at the national boundary of a nation. Thus, there is continuing and growing need to coordinate and schedule
various uses among nationals in order to reduce
interference to an acceptable level. Furthermore,
some technological development and utilization involve international uses and adoption. Such technological developments rely directly on international diplomacy.

Several basic points about the international
dimension should be made at the outset. First,
much of the international interaction involving
problems of telecommunications is based on the pur-
suit of national goals of foreign policy. Various
states seek allocations and other agreements which
benefit their own users or their own uses. However,
in apparent contradiction to this statement, there
are many examples in which the participants seek
to reach mutual accommodations for the use of spectrum or facilities. Thus, there is an element of
tolerance and accommodation among national users
of telecommunications facilities so that all comers
may receive some benefit. In addition, there are
certain elements of squatter's rights in spectrum
allocations. This unwritten rule of registration
of spectrum uses--the first come, first served

principle--is evident in various, competing claims
for the use of spectrum. Lastly, it is possible
to use the airwaves and other telecommunications
systems for political propaganda purposes. These
political objectives might be rejected or opposed
by various groups of users, but it is likely that
technology will be developed and used for such
purposes.

International Institutions

Over the years ad hoc, international bodies have
been developed for dealing with specific telecommunications problems. However, these are crea-
tures for an immediate problem and usually do not
last beyond the problem which gave them impetus.
Also, there are permanent bodies such as the International Telecommunication Union (ITU), a
special, permanent United Nations agency which
deals on a continuing basis with various international communications problems. The ITU provides
the central body for clearance and registration of
spectrum-use requests made by member nations. The
United Nations itself has been dealing with some
telecommunications issues on a continual basis
through its ad hoc Committee on the Peaceful Uses
of Outer Space. This body prepares studies and
makes general policy recommendations, such as
communications satellites, which relate to international uses of space.
    As an example of a permanent international body
in this field, the International Frequency Registration Board (IFRB) is a special agency within
the ITU which handles the registration of spectrum
uses by nations.(25) Although the uses to which
nations put the spectrum are supposedly noninter-
fering, and the IFRB supposedly does have enforcement authority, the general rule of operation for
the board is to accept any registration on a first
come, first served basis, with little enforcement
of regulations regarding usage. Some would argue

that such an ineffective policing body serves no
purpose and should be abolished. However, the
mere function of registration is important, because it provides a central clearing mechanism
for recording uses and allocations. The utility
of this depends on the objectives a nation is
pursuing, and the effect of the process on the
achievement of those goals. A nation is likely to
support the IFRB process if that country has been
able to secure registration of its uses first, and
thus enjoy some protection by the IFRB. Another
nation, a late comer in registration, may oppose
the process because it has been frozen out of a
particular portion of the spectrum.

In addition to the IFRB, the ITU has two other
units which deal with telecommunications problems
at the international level: the Consultative Committee on International Radio (CCIR) and the Consultative Committee on International Telegraphy
and Telephony (CCITT). The CCIR is concerned with
standards for various aspects of long distance
radio communications. This function overlaps the
IFRB, except that the IFRB deals primarily with
the problems of radio interference and registration, while the CCIR focuses on other standardization matters. The CCITT devotes its time largely
to making studies and recommendations on a wide
number of aspects of telephony and telegraphy.
This organization uses study groups to reach various conclusions and standardization recommendations on a vast array of telecommunications topics
such as switching, signaling, noise, transmission,
and performance.

These three bodies illustrate the value of international cooperation and its tentativeness.
Many more problems would arise in international
communication if it were not for these clearing
agencies. However, their success depends on the
willingness of members to cooperate and accept
various policies. Usually, the bodies will not

make a recommendation without careful study and
widespread support from participants. This insures the adoption of the recommendation and the
continued success of the body. These bodies deal
with matters about which there may be little,
heated political conflict, and this increases
their viability. On the less widely accepted ad
hoc issues, such as the INTELSAT arrangements and
operations, national competition still arises, and
may interfere with the technology and its utilization.

INTELSAT is another international body which
directly relates to an important area of international telecommunications. Recently reformulated
as a permanent body with operating procedures,
INTELSAT is the collection of nations which jointly own and operate the international communications satellite system developed since the 1962
creation of the United States Company, Communications Satellite Corporation (COMSAT).(26) The creation and operation of INTELSAT is an interesting
example of the mixture of national and international politics. Most importantly, it indicates
the roles of technologically dependent and independent countries in terms of international cooperation. Given the technological superiority of
the United States in communications satellite development since 1958, this country pursued a set
of goals for the development and utilization, as
well as control, of such a particular communications system. The technological dependent countries of Western Europe were in a weak bargaining
position during the early phases of INTELSAT's
operation. Only in the negotiations for the
permanent arrangements of INTELSAT, during the
1969 to 1972 period, did these countries exert
some significant pressure that adjusted the arrangement for operation of the agency and procurement of the satellite technology.

INTELSAT, both the temporary and the permanent
structure, illustrates a number of policy preferences. Various participants sought to foster their
own objectives or obtain various benefits from the
arrangement. Whether each country achieved what
it wanted or is satisfied with the outcome is
another question. It is important to consider that
the development of this particular technology and
its use for international communication depends
upon international politics and various foreign
policies which may be at odds with one another.

The International Actors

Certainly the most prominent actors at this level
are the nations which have some interest or economic stake in international communications. Some
nations are users only, while others not only
use the service but also provide the technology
that makes these communications systems possible.
While it is not always possible to discuss a *national* interest, since there may be competing interests within a country, one can view the state
or official governmental representative as the
primary actor in the international arenas. Thus,
the foreign policy objectives of individual countries may become important aspects of international
telecommunications matters, just as the technology
is only one part of the domestic technology development of telecommunications. The conflicting
national objectives which arise are interesting,
and they tell a good deal about what agreements
and other results appear from international negotiations.

Among the countries which operate in the international setting, an important distinction can be
made between those which have the technology and
seek to "export" it, and those which are technologically "poor" and are recipients of the technology. There are interesting cases in which the
exporting nation has made major, overt efforts to

pursue various policies which benefit it, while
imposing costs and other limitations on the target
countries. Generally, the advanced countries have
the technological capability to export devices and
services, while the developing countries tend to
be the target, users of various devices. Only the
United States and Russia were in the position of
exporting satellite communications during the mid-
1960's and, as a result, could nearly dictate the
terms under which they provided the service and
the technology. Although some European countries
have worked vigorously on developing an independent technological base for such services, they are
limited by the difficulty in launching communications satellites without an active space program.
Some European countries have even united in an effort to develop an independent launch capability
as well as communications satellite ability. Their
efforts have enjoyed only limited success to date,
facing a good deal of political and economic opposition from the established exporter (the United
States).(26)

The developing countries or any recipient country may not readily accept a technology or a service which is being exported to them. The inde-
pendent/dependent relationship of these countries
tends to make them suspicious and often antagonis-
tic to proposals from a dominant country. Even if
an important, revolutionary technology would
greatly facilitate various kinds of international
communications, its utilization may be hampered or
even blocked by national jealousies and competition.

In addition to the national actors, some users
of technology and communications channels are
vitally concerned with international policies regarding these areas. Thus, the international carriers who provide international links for voice
and record communications are centrally interested
in the policies operating among countries or with-

in international organizations which affect them.
In addition, the individual who uses international
communications channels may be affected by these
policies. However, rarely are individuals capable
of having much impact on such international policies. Beyond the benefit of reduced rates for
international communications, these individual
users have little direct concern.
What this means is that even on the international level, possibly even more so there than at
the domestic level, the adoption and use of technologies may be hindered by nontechnical factors.
The competition between cultures and political
beliefs may increase the problems of technological
adoption. The competition among nations for prestige and other nonmaterial benefits, as well as
the real competition in terms of economics and
political objectives, may shape the development
and use of technology.

A General Model

Figure 3-1 diagrams the kinds of relationships
which can exist in the domestic political process,
with particular reference to the telecommunications
industry. It shows the actors who most frequently
appear in the process, the arenas in which the decisions are made, and the potential interactions
between the interests and the arenas. It is interesting to note that this general model has outlined
all the possible combinations of interests and institutions. In this form, the model is not very
helpful to the observer who is interested in the
particular relationships for a specific policy decision. This model, throughout the discussions of
case studies which follow in subsequent chapters,
will delineate interests and arenas in each particular decision and show the relationships which
existed in that case. In that form, the model will
summarize the allocation process as it occurred in

each case. The various lines, specified in the key
to Figure 3-1, will be used in later figures to indicate the strength and direction of various relationships in the particular policy area being ana-
lyzed.

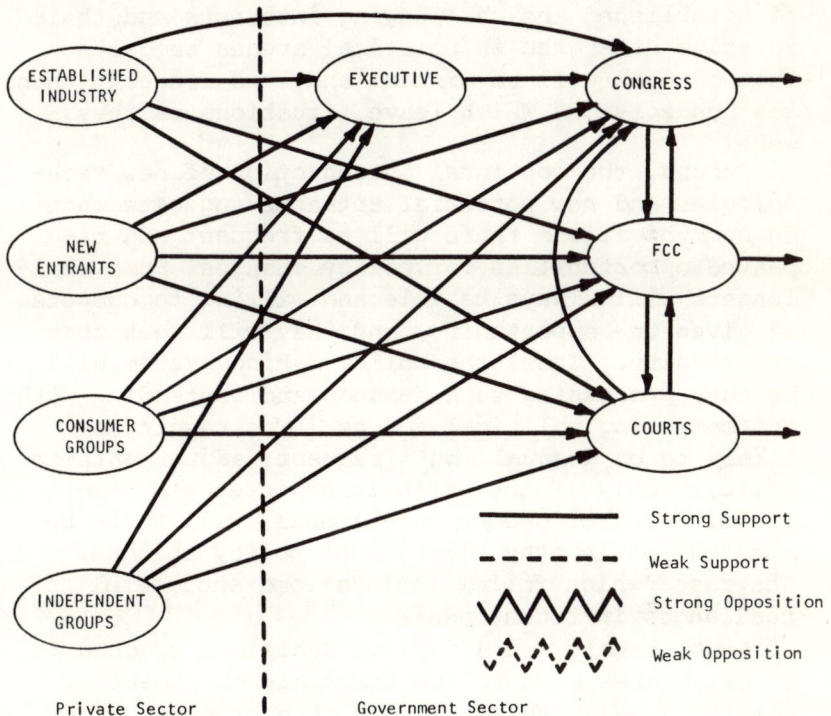

Figure 3-1 General technopolitical allocation model

It is difficult to present causal relationships
for any case since "causation" in the social sci-
ences is difficult to establish. However, the
models which follow for the cases are qualitative
evaluations of the actors and arenas and results
which appear to have been controlling. Hopefully,
future investigations of these and other policy
examples will result in more precise and certain
statements about causation.

Several points can be made about the policy results of this model and discussion.

First, most policies are not likely to break
drastically with established policy patterns.(27)
Rather, any new policy probably will be an incremental change from the status quo. The balancing
of established and challenging interests and their
relative strengths in political arenas suggests
that changes will be minimal and less frequent than
the nondecisions which leave situations as they
were.

Second, the continual introduction of new technologies and new potential entrants suggests that
in communications there will be frequent and repeated opportunities for policy changes. Some challengers will always be able and willing to compete
if given the opportunity, and they will seek that
opportunity. Thus, the policy-making system will
be busy processing such demands and contending with
the competing political forces. The result is
likely to be gradual, but frequent, adjustments in
policy. Only if the established interests completely control one or more arenas, will there be
a major and lengthy blockage of policy changes.
The cases which follow include some successful
challenges in recent years.

Last, rarely are any final decisions reached in
a policy area. There are temporary or momentary
results upon which the actors then proceed to develop technologies and provide services. However,
most policy decisions are open to subsequent revision, as actors or forces lose or gain strength
in various political arenas. Thus, the cases explored in this book indicate temporary resolutions
of policy questions, and in some cases these have
been modified or reversed since the time of the
decision. The lack of final results indicates that
public policies are never fixed, and resource allocation by these processes involves a continual,
ongoing contending of various interests, rather

than final outcomes of the competition among interests.

The Economics of Telecommunications

There are two major parts of the telecommunications
industry, and these are treated separately in chapters to follow. The broadcasting industry utilizes
a particular set of technologies in order to reach
the largest possible audience, while the common
carrier industry provides services to any user on
an individual use and cost basis. These two kinds
of services are reflected in various economic considerations, and they structure the use of technology and the services which are available to
whatever kind of user.

The economic considerations discussed here are
intended to provide a general perspective which
will be developed in more detail later, when dis-
cussing particular decisions. However, it should
be apparent that the economics of telecommunications do not operate with theoretically perfect
efficiency. Rather, market imperfections, government intervention and allocation, and various political constraints add to the "economic" considerations a substantial amount of imperfection and a
need for various kinds of artificial structures on
the economic functioning of the industries involved.
Our discussion is divided into the two separate
areas of telecommunications since some of the characteristics of each are unique. It is designed to
provide an appreciation of the general features of
the economic setting of telecommunications.

Broadcasting Industry

The structure of this industry is based on broadcasting a single program from a single source to
as large an audience as can be attracted. In this
type of commercial venture the amount of revenue

a broadcaster receives from an advertiser is based
on the size of the viewing or listening audience
the broadcast reaches. In turn, the size of the
audience depends on the attractiveness of the program being broadcast. So, it is to the advantage
of the broadcaster to present programming which
will attract as large a commercial audience as possible. The revenue from such broadcasting is derived from commercial advertisers who "buy" advertising time to sponsor the program, depending on
the size of the audience and the time of day during
which the program is aired.(28)

The logical conclusion of this structural feature would be a single program which would attract
all the viewers. However, this extreme has not
materialized, although many critics of present-day
broadcasting complain about the dearth of program
diversity. In addition to this program concentra-
tion feature, based on attracting commercial viewing audiences, the costs of program distribution
have led to the development of networks which offer
a complete set of programs to affiliate stations in
an effort to: (1) reduce the difficulty of obtaining large, nationwide audiences for advertising
purposes; (2) centralize the acquisition of high-
quality and entertaining programs from a diverse
set of program producers; and (3) make the local
broadcaster's need for program production minimal.
The networks do not produce programs but rather
purchase these from program producers, which include the Hollywood movie companies among others.
Then the networks offer the individual broadcasting
affiliates the option of a complete package of programs.

The centralizing tendency of the advertising and
the programming features of broadcasting conflict
with a basic policy followed for decades by the
FCC. The commission has operated with the objective of providing a locality (the smallest possible
audience segment) with at least one local source

of programming. That means that the FCC seeks to
foster the development and operation of local
broadcasters (local radio and television stations)
under the assumption that such local operation will
provide the citizens with local programming and
other features which are of local, rather than
national, interest. Thus, the allocation of the
broadcasting spectrum among broadcasters is based
upon this policy assumption and objective, and is
in conflict with the actual economics of commercial
broadcasting.

The basic reason for licensing of broadcasters
is the scarcity of usable spectrum for these purposes and the early congestion which resulted from
overuse of the spectrum by broadcasters in the
1920's.(29) This policy will be discussed in greater detail in Chapter 4. It should be apparent, however, that the need to allocate the scarce resource
of electromagnetic spectrum on the basis of some
policy--in this case, localism--has produced a
countervailing, if not successful, balance to the
centralizing tendencies of program production and
commercial audience needs. Whether the broadcast
industry would operate in the same economic fashion
if the policy of localism did not exist is an open
question. It appears that the strong pressures of
the network and the commercial advertising objective of mass audiences, rather than isolated, local
audiences, would only be more pronounced than they
are at this point, if the regulator were following
a nonlocal licensing policy.

Licensing is the FCC's major power over the
broadcasting industry. The requirements that the
license be renewed every three years, that no
broadcaster can trade in licenses (trafficking) and
that no broadcaster obtains a property right in the
license clearly is designed to indicate that the
"airwaves belong to the people" and they are being
loaned, with no charge to the broadcaster, for use
in the "public interest." Regardless of this

statutory view of the license to broadcast, the
broadcasters do sell stations for much more than
their market value, and this additional revenue is
often seen as the value of the license that goes
along with the equipment and the building.(30) This
is especially true of broadcasting stations in
large viewing areas--large cities or metropolitan
areas--since these stations reach large commercial
markets and thus generate substantial advertising
revenues to the station owner.

In this economic situation, there is no guarantee that the market--commercial advertising to mass
audiences--provides any clear indication of program
quality or approval as a measure of the public interest. Some observers have suggested that a
system, such as subscription or Pay TV, which permits the viewer to "vote" for programs by directly
paying for them would be more economically efficient. This policy area is developing and will be
discussed in more detail in later chapters. However, if such direct payment for programs were allowed, the "middle man"--networks, and commercial
advertisers--would be eliminated. Broadcasters and
viewers would have to bid for attractive programs
with the program producers. A large portion of the
existing broadcasting industry opposes this kind of
mechanism because of the possible and profound ramifications it would have for economic status quo.

Common Carriers

There are two justifications for regulating common
carriers by a governmental agency such as the FCC.
The first is identical to the regulation of
broadcasting--the need to allocate the scarce
electromagnetic spectrum. The frequencies and
other technical constraints involved in common
carrier use of the spectrum require close supervision and regulation.
The second reason for common carrier regulations
is that natural monopolies develop in this area,

largely due to economies of scale and the capital intensive nature of the industry, so that government regulation is necessary to protect the public from the abuses a monopolist (single supplier of a good) could impose on the consumer.(31)

There has been recent debate over whether such justification warrants regulation. For example, as technology develops, it is possible that common carriers will not have to use the electromagnetic spectrum to provide their services. Laser beams, fiber optics, and waveguides, which may be the next stages of technology in this field, do not use the radio spectrum, and they provide little potential interference with other uses of the spectrum, and thus regulation is not needed. That may not preclude government regulation, since the FCC has authority to regulate "wire" communication, but it does remove one of the justifications for such regulation.

Some observers have attacked the natural monopoly idea because it is unclear that regulation of natural monopolies produces the effects that are anticipated.(21,31) Furthermore, even the FCC has, in recent years, sought to inject certain elements of competition into the common carrier industry as a means of regulating or controlling the carriers. This will be discussed later. Whether competition will be successful in regulating or in achieving the other objectives of the policy makers remains to be seen. However, it does suggest that alternatives to the regulatory framework established and developed by the FCC for the past four decades might be viable.

The economies of scale and the capital intensive nature of common carriers do present major economic considerations to the problem of common carrier regulations.(32) It is not at all clear that any new common carrier can compete successfully with the established carriers because of the costs of building a competing system of carriers and because the

economic success of carriers depends on providing
large varieties of services to large numbers of
users before common carrier operation becomes profitable. Selective competition by new entrants
might be successful so long as the existing carriers are not allowed to "compete" for the same
business as these specialized carriers. This will
be discussed in Chapters 5 and 7. However, whether
this is an appropriate policy for the government to
pursue and whether it will be successful in producing the kinds of results anticipated is still an
open policy question.

Conclusions

This discussion has outlined some of the important
political and economic factors which structure the
development and use of telecommunications technologies and services. The specific political and
economic considerations in any particular case will
vary markedly with the details of the situation.
The following chapters discuss a number of specific
decisions by various government agencies and actors
and illustrate the effect and impact of various of
these political and economic considerations on
telecommunications. That discussion should also
provide a sound basis for estimating likely developments regarding current policy questions in the
telecommunications industry.

Chapter 4

Broadcast Communications

Of all telecommunications services the most widely
used are those of radio and television broadcasting. These services and the political arena in
which they exist provide a wide range of examples
of the interaction of technology, politics, and society. The major story of the development of broadcast technology is tied to the FCC's efforts to
allocate spectrum and regulate broadcast communication. Several examples of policies which have
been formulated to use the spectrum efficiently
are explored in the following sections. They illustrate some of the political and social forces
which affect the use and development of broadcast
technology.

Several factors about broadcast communication,
which differentiate it from other kinds, should be
kept in mind.

First, broadcasting is a one-way process in
which the broadcaster communicates with listeners,
and the listener normally does not respond directly
to the communication. At most, the listener may
respond through a commercial transaction such as
buying the advertised product presented as part of
the broadcast message. Although recently many
broadcasters have instituted talk-back programs,
the basic function of broadcasting is one-way.

The second major characteristic of broadcast
communication is that it is aimed at mass audiences. Instead of communicating with a particular

individual, this form of communication seeks to
reach as large an audience as possible, so that
the commercial messages presented will reach the
largest potential market.(1) The financial success
of broadcasters depends on the advertising revenue
they attract. This revenue, in turn, depends upon
the number of listeners or viewers who tune to a
particular broadcast and thereby become targets
for the commercial message. This focus has a significant impact on the shape of programming and on
the economics of the industry. These matters are
discussed more extensively later.

The Broadcast Spectrum

Allocation Procedures

Several factors have strongly influenced the development of radio broadcasting. Mostly, these
factors relate to the nature of the electromagnetic
(radio) spectrum and the process by which it is allocated. The electromagnetic spectrum provides
the communication channels through which broadcast
messages are disseminated in this country. The
general nature and technical constraints of the
spectrum are discussed in Chapter 2. One important
fact about the spectrum is that it can be viewed
as a "free" resource only in a limited fashion.(2)
Although there is no spectrum market, and the users
of the spectrum (broadcasters) acquire no property
right to their assigned frequencies, the spectrum
can be polluted and so overused that no user enjoys
an unhampered portion of it. The spectrum can become overcrowded if too many users seek to broadcast on the same frequency or if the uses of it
conflict. The basic outline which the regulators
have followed to reduce such congestion is to space
users geographically, by time and by frequency.(3)
Thus, if two broadcasters use the same frequency
but are sufficiently far apart, their use will not

interfere with each other's broadcasts. No interference occurs if the broadcasters do not use the
frequency at the same time. Lastly, two broadcasters would not interfere with one another if
they broadcast in different portions of the spectrum.

Most of the allocation process is focused on
this need to accommodate various demands in a
fashion that does not produce intolerable amounts
of interference among users. The market clearing
processes which would allocate the spectrum on the
basis of price and which would assess the costs
to the appropriate users has been replaced. A
set of noneconomic processes has been established
which influences the uses and the amount of spectrum occupied by a variety of broadcast users. The
major costs (which must still be considered) relate
largely to the political, regulatory allocation
process. Thus, the way in which one insures that
his uses of the spectrum are not interfered with
is through political lobbying so that the allocator will not permit interference with the existing users' allocations. This political process
and the accompanying costs may involve lobbying
before Congress with regard to a proposed statute
which would alter the allocation process or the
criteria used for allocating. It may require that
the broadcaster seek special or individual protection from the Executive Branch. Most frequently
the allocation process for broadcasters requires
the continual costs of protecting the broadcaster's
vested interests in a portion of the spectrum.

The political costs of this process may be very
expensive. However, this particular procedure is
used because various observers and actors, including broadcasters, feel it is advantageous over the
free-market process of allocation. While there is
much debate about an allocation process which depends on political lobbying rather than economic
criteria for success, the facts are that the FCC

has statutory authority to allocate portions of
the spectrum for private users. The "public convenience, interest, or necessity" goals of this
process have been left to the agency to define as
it chooses.

Governmental use of the spectrum is not regulated this way, and while the government users are
supposed to coordinate their use of spectrum with
others, government is not constrained by any statutory requirement to accommodate other users.(4)
There are examples of government interference with
private users which have even resulted in government users excluding private users from certain
portions of the spectrum.(5) Usually, the usage of
spectrum is coordinated between government and private users, and a significant portion of the government's spectrum allocations overlaps private
allocations without any great interference or conflict between the two groups. Yet the precedent
has evolved of giving the government what it wants
on the basis of its asking for it.(3)

Several general characteristics of the allocation system merit attention. The development of
technology for broadcasting may depend largely on
what uses are being made of which portions of the
spectrum. That is, if there is ample bandwidth to
accommodate a current use, there may be no incentive to innovate or use different technologies.
But, if there is much congestion from overcrowding,
one of two kinds of technological changes may occur. New technologies may be developed at the intensive margin to permit more users to occupy the
existing allocation with less or no greater interference and congestion than existed before. If
such innovation is not possible, the extensive
margin may see an innovative effort. Thus, if
there is no room in the existing band, technology
may be forced to develop higher portions of the
spectrum and adapt it for the original use. Both
intensive and extensive margin development are well

documented, and they illustrate two technological responses to the same congestion problem.(2,3)

The initial way in which spectrum was allocated was on a first come, first served basis. As a new use arose, requiring further allocation of spectrum, the allocator simply assigned it a portion of the unoccupied spectrum, usually in the upper ranges of the then usable spectrum. Initially, this was done with little examination of the relative values of old and new uses of the frequencies. Generally, this allocation process occurred when a large amount of spectrum was unused and pressure from users was minimal. Thus, the newer entrants were assigned higher segments of the spectrum and the older established user occupied the lower portions of the spectrum. The difficulty with this procedure is that all portions of the spectrum are not equally attractive and feasible for all kinds of uses. Certain kinds of broadcasting may be done more effectively at lower frequencies, but they may be allocated a higher set of frequencies simply because their requests were made late.

This feature of the allocation process in the United States means that a de facto form of squatter's rights has developed. Under such a scheme, the first to seek often obtained permission to use a particular portion of the spectrum. Potential users who sought authorization for the same frequency carried the burden of dislodging the occupant. This situation is particularly important in terms of technological innovations. First, it causes new technological innovations to be directed toward the upper, unused, portions of the spectrum. Second, new services requiring utilization of the presently occupied portion of the spectrum also required strong political efforts as well as technological innovation.

Frequency Allocation Table

The FCC has developed a number of mechanisms and
procedures to permit long-term as well as short-
term (first come, first served) allocation requests. These are efforts at reserving portions
of the spectrum for a particular use before a
request is presented. A major means of doing this
is the frequency allocation table which has been
used for FM radio broadcasting and for VHF and UHF
television broadcasting. The allocation table is
essentially a block of frequencies reserved for
particular uses. The block takes into account the
amount of bandwidth required for a single broadcaster to have an interference-free operation. The
block also reflects the FCC's general policy outlines about a particular industry. That is, the
total amount of the spectrum devoted to a particular function will reflect how important the FCC
considers the development of that spectrum use to
be. Thus, the allocation for UHF television, which
is quite large, reflects a major FCC emphasis on
encouraging certain kinds of television development
with many UHF television channels.

In fact, the allocation table may allocate large
portions of the spectrum, currently empty, and it
may prevent use of those frequencies, pending the
emergence of the designated kinds of users. The
table reflects established policies for allocation
of frequency, and in that way the future allocation to users is insured by the original determination and establishment of the allocation table.
A portion of the current FCC Allocation Table appears in Table 4-1. The amount of bandwidth required and the number of users determine the amount
of spectrum allocated for a particular service. For
example, the 70 UHF TV channels occupy a 420 MHz
segment of the spectrum. In addition to bandwidth
and number of users, the spectral location of segments reserved by the allocation table can strongly
influence the development. Thus, moving the FM

frequencies from their original location to a
higher portion of the spectrum in 1945, contributed
significantly to retardation of the FM industry's
development.(6)

On occasion a user has the opportunity to stock-
pile spectrum even though he will not use it immediately, if ever. This stockpiling may be done
to exclude competing entrants from using spectrum
or simply to horde spectrum. This has been a major
criticism of government use relating to national
security matters.(3) While the allocation table reserves spectrum for a particular use, stockpiling
reserves spectrum for a particular user. There are
not numerous examples of stockpiling.

In contrast to stockpiling, some portions of the
spectrum are allocated to a particular use, and
anyone interested in that use, who has the necessary equipment, can obtain a permit. The resulting
congestion and interference in these portions of
the spectrum are notorious; a prime example is
citizens band (CB) radio. Users of this service
essentially have a "party line" and anyone with the
equipment and the permit can use the frequency
whenever he or she wishes and whenever the interference level is low enough for his or her purpose.

The allocation of radio spectrum to users depends on a variety of factors which relate to
economics and politics as well as to the technology
involved in various uses. Broadcasting use of the
spectrum has built-in incentives to protect the
status quo once arrived at. That a new entrant or
a new user of frequency can easily dislodge an
established user is unlikely. Thus, by means of
allocation tables, which segregate large portions
of the spectrum for users and future users, and the
ad hoc, squatter's rights principle which relates
to other portions of spectrum usage, there is no
easy way that a new technology or technological innovation can be rapidly assimilated into the broadcasting system.

Table 4-1 Example of FCC table of frequency allocations
Source: 47 C.F.R. § 2.106 (1976).

§ 2.106 Table of Frequency Allocations—Continued

Worldwide		Region 2		United States		Federal Communications Commission				
Band (MHz)	Service	Band (MHz)	Service	Band (MHz)	Allocation	Band (MHz)	Service	Class of station	Frequency (MHz)	Nature of services or stations
1	2	3	4	5	6	7	8	9	10	11
				37-38	NG.	37-37.01	LAND MOBILE.	Base. Land mobile.		INDUSTRIAL.
						37.01-37.43	LAND MOBILE.	Base. Land mobile.		PUBLIC SAFETY.
27.75-28.25	FIXED. (228) (229) MOBILE. (231) Radio astronomy. (233B)					37.43-37.89	LAND MOBILE. (NG49)	Base. Land mobile.		INDUSTRIAL.
				38-39	G. (US41)	37.89-38	LAND MOBILE.	Base. Land mobile.		PUBLIC SAFETY.
28.25-41 (235) (236) (236A)	FIXED. (230) (231) MOBILE.			39-40	NG. (US94)	39-40	LAND MOBILE.	Base. Land mobile.		PUBLIC SAFETY.
				40-42 (US220) (226)	G. (US94) (US210) (US230)				40.68	Industrial scientific and medical equipment.
41-50		41-50	FIXED. (228) (231) MOBILE. (233A) (236A) (237)	42-46.6	NG.	42.95-43.19	LAND MOBILE.	Base. Land mobile.		PUBLIC SAFETY.
						43.19-43.69	LAND MOBILE.	Base. Land mobile.		INDUSTRIAL.
						43.69-44.61	LAND MOBILE.	Base. Land mobile.		DOMESTIC PUBLIC. INDUSTRIAL. PUBLIC SAFETY.
						44.61-46.6	LAND MOBILE.	Base. Land mobile.		LAND TRANSPORTATION.
				46.6-47	G.					
				47-49.6	NG.	47-47.43	LAND MOBILE.	Base. Land mobile.		PUBLIC SAFETY.
						47.43-47.69	LAND MOBILE.	Base. Land mobile.		PUBLIC SAFETY. INDUSTRIAL.
				49.6-50	G.	47.69-49.6	LAND MOBILE.	Base. Land mobile.		INDUSTRIAL.
50-54		50-54	AMATEUR.	50-54	AMA-TEUR. (US1)	50-54	AMATEUR.	Amateur.		AMATEUR.

Band (MHz)	Allocation	Band (MHz)		Band (MHz)	Allocation	U.S. service	Frequency / service
54-68	FIXED, MOBILE, BROADCASTING. (250)(287)	54-72 (U830)	NG.	54-72 (NG56)	BROADCASTING.	Television broadcasting.	44.25 Video } Channel 2 49.75 Sound } 61.25 Video } Channel 3 67.75 Sound } 71.75 Video } Channel 4 Sound }
68-72.0	FIXED, MOBILE, BROADCASTING.	72-73 (U830)	NG.	72-73 (NG56)	FIXED. (NG1)(NG3)(NG49)	Operational fixed.	72.02-72.98 (NG33) Operational fixed.
72-74.6	RADIO ASTRONOMY. (253A)(253B)	73-74.6	G, NG. (U821)(U8100)	73-74.6	RADIO ASTRONOMY. (U574)	Radio astronomy.	75 RADIO ASTRONOMY.
74.6-75.4	AERONAUTICAL RADIONAVIGATION.	74.6-75.4	G, NG.	74.6-75.4	AERONAUTICAL RADIONAVIGATION.	Aeronautical radionavigation.	Marker beacon.
75.4-88	FIXED, MOBILE, BROADCASTING.	75.4-76 (U830)	NG.	75.4-76 (NG56)	FIXED. (NG1)(NG49)	Operational fixed.	75.4-76.98 (NG33) Operational fixed. 77.25 Video } Channel 5 81.75 Sound } 83.25 Video } Channel 6 87.75 Sound }
88-100	BROADCASTING.	76-88 (U827)	NG.	76-88 (NG21)	BROADCASTING.	Television broadcasting.	88.1-107.9 (NG36) FM Channel 201-FM Channel 300.
100-108	BROADCASTING.	88-108 (U823)(U893)	NG.	88-108 (NG21)	BROADCASTING.	FM broadcasting. (NG2).	
108-117.975 AERONAUTICAL RADIONAVIGATION.		108-117.975 (U893)	G, NG.	108-117.975	AERONAUTICAL RADIONAVIGATION.	Radionavigation land.	108.05 Omnidirectional range (VOR). 108.10 Localizer. 108.15 do. 108.20 Omnidirectional range (VOR). 108.25 do. 108.30 Localizer. 108.35 do. 108.40 Omnidirectional range (VOR). 108.45 do. 108.50 Localizer. 108.55 do. 108.60 Omnidirectional range (VOR). 108.65 do. 108.70 Localizer. 108.75 do. 108.80 Omnidirectional range (VOR). 108.85 do. 108.90 Localizer. 108.95 do. 109.00 Omnidirectional range (VOR). 109.05 do. 109.10 Localizer. 109.15 do. 109.20 Omnidirectional range (VOR). 109.25 do. 109.30 Localizer. 109.35 do. 109.40 Omnidirectional range (VOR). 109.45 do. 109.50 Localizer. 109.55 do. 109.60 Omnidirectional range (VOR). 109.65 Do.

See footnotes at end of table.

Constraints on Broadcast Program Content

The content of broadcasters' programs is an important aspect of communication. In theory, the free
speech and free press provisions of the First
Amendment of the U.S. Constitution protect broadcasters as well as the more traditional forms of
expression. Interestingly, there are several means
by which the regulatory powers of the FCC have
been used to inject certain, minimum standards
which broadcasters must meet in order to satisfy
the statutory requirements of licensing. These requirements are based on the maxim that a license
holder must perform to meet the public interest.
The judge of whether the broadcaster has met the
public interest is the FCC and this is done when
a license is renewed every three years. In this
way the FCC has gradually established a set of
criteria for judging whether the public interest
has been served by the kind of programming which
the broadcaster has presented or proposes to present under the license. Few denials of licenses
or renewals have been based on the kinds of programming offered, but that may be because most
broadcasters have met the minimal FCC requirements.(7)
    Even in the absence of formal program controls,
certain economic considerations structure the content of programs that mass communication broadcasters present.(1) The need to attract the largest
audience possible requires that the programming
attract the viewer or listener. That means that
the programming must be either unique, if the advertiser is seeking to reach a particular and specialized audience, or, more likely, that it be
attractive to the largest possible set of buyers.
The attraction of the largest audience possible
tends to prevent use of the mass media for particularized or unique audiences, and it dictates

that the programs presented be widely acceptable
in content and tone.

The fear of audience "fractionation" which
would splinter audiences into particular groups
has long been an argument by some against certain
kinds of technology, such as cable TV, which would
facilitate particularization of audiences. Audience fractionation has become a common phenomenon
for AM radio which now presents rather narrow and
specialized programming. Mass audience, TV network
programming, based on the delivery of large audiences to sponsors, could not function profitably
as it does now with specialized programming or
fractionated audiences. While one might argue that
specialized programming would improve the "quality"
of the programs offered, quality is a subjective
evaluation. Many observers argue that the market
determines the program content currently, and
nothing should be allowed to change that--whether
it be programming requirements or technological
innovations which make fractionation possible and
profitable.

Examples of Broadcast Technology Development

Discussions to follow illustrate some ways in which
technological development and usage in the broadcasting industry have been encouraged or hindered
by policies. In particular, the FCC has sought to
achieve a variety of explicit and implicit policy
objectives with regard to technology by pursuing
various policies over particular spectrum allocation questions. The kinds of issues they sought
to address, the means they chose to achieve their
particular ends, and the outcomes of these efforts
will be outlined.

UHF Television

In 1962 Congress enacted the All-Channel Receiver
law which gave the FCC statutory authority to re-

quire that all television sets sold in interstate
commerce be capable of receiving all television
channels. (See Appendix).(6,8) This was not an
isolated piece of legislation but rather an overt
attempt by the FCC to foster the use of UHF television channels which it had made available as
part of the original allocation table for television in 1952. In the 1952 allocation, the FCC announced that it sought to encourage the development
of diversity of programming for TV by increasing
the number of stations on the air and to provide
the maximum number of local television stations.(9)
To do that, the VHF spectrum was inadequate, so a
substantial portion of the UHF frequency was allocated for some 70 television channels. In doing
this, the FCC increased the number of channels
from 12 VHF to a total of 82.

These developments to encourage diversity and
the proliferation of stations and programming by
increasing the channels available may not be sufficient to produce the results the FCC sought. As
the data presented in Figure 4-1 indicate, the
number of UHF stations on the air, compared with
VHF stations, shows that the growth of UHF stations
has been slow, despite the fact that the number of
potential UHF assignments is much higher than VHF.

A variety of economic and technological factors
accounts for this failure of UHF television to develop.

First, technologically, the UHF portion of the
spectrum is not as good a media as the VHF spectrum
for broadcasting television. The UHF broadcast of
television is subject to a number of local interference problems, such as tall buildings and other
structures. Furthermore, energy radiated at UHF
wavelengths attenuates much more rapidly than does
energy at VHF wavelengths, so UHF signals do not
carry as far as VHF signals. Thus, the UHF spectrum is not as attractive as is VHF spectrum for
the TV broadcaster. Audience size is generally

smaller for UHF because of smaller viewing area and
a frequently poorer picture quality. The smaller
audience sizes mean that the advertising income for
a UHF station will be less than for a comparable
VHF station with identical programming.

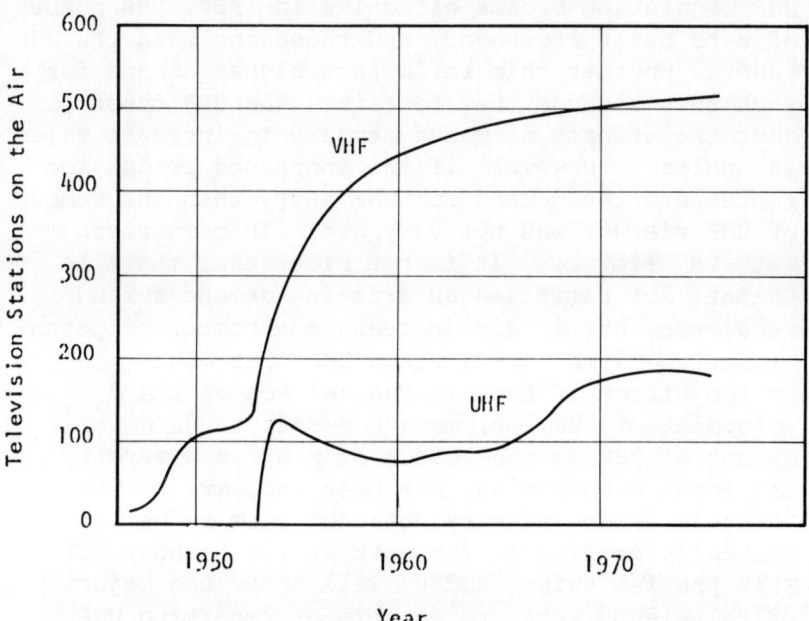

Figure 4-1   Television stations on the air in the
             United States, by year. Source: *Television
             Factbook,* Volume 46, 1977 edition.

A second important constraint on the development
of UHF audiences in the 1950's was the simple fact
that the standard television receiver could not
receive UHF signals. In the 1950's as television
was developing, manufacturers produced sets which
would only receive signals then on the air, and
there were VHF only. To receive UHF signals in the
1950's, a viewer not only had to opt for a poorer
quality of picture, but also had to purchase a

separate converter or a special television set.
Given the fact that most broadcasts in this period
were VHF, it was difficult to find much enthusiasm
for UHF TV. The All-Channel Act of 1962 was an
obvious effort to increase the availability of UHF
receivers. Figure 4-2 indicates that, when the
UHF regulation became effective in 1964, the number
of sets built increased, and these included the UHF
tuner. Whether this reflects a higher demand for
such sets because they contained the UHF tuner or
just the efforts of manufacturers to increase sales
is unclear. However, if the increased production
represents the demand for UHF sets, then the number
of UHF viewers was not very great in comparison
with VHF viewers. It is not clear that the All-
Channel Act satisfied an existing demand for UHF
receivers, but it did increase the number of potential viewers who could watch UHF stations.

The effect of the All-Channel Act on the development of UHF, or, more generally, the development of UHF as the source of program diversity
and local programming, has been unclear.(9) The
UHF television industry does not seem to be economically profitable for most of its members.(2)
This problem exists today, well after the majority
of television sets are capable of receiving UHF
signals along with VHF and well after many FCC
efforts to encourage development of UHF. In fact,
many UHF stations are not profitable, and, even
among those which are profitable, the amount of
profit is much lower than comparable VHF stations.
(2) Many reasons for this lack of UHF success and
the under utilization of the spectrum allocated
for this use have been suggested.

Technical problems still exist and over-the-air
UHF broadcasts frequently are not of the same quality as VHF broadcasts. Possibly, if sufficient
economic or other incentives were presented, new
technologies could produce a better UHF signal, but
such a development appears unlikely. Recently, the

FCC has required that a detent or "click" tuner be
incorporated into all UHF receivers.(10) This requirement will make selection of UHF channels easier, but the impact of this requirement cannot be
predicted.

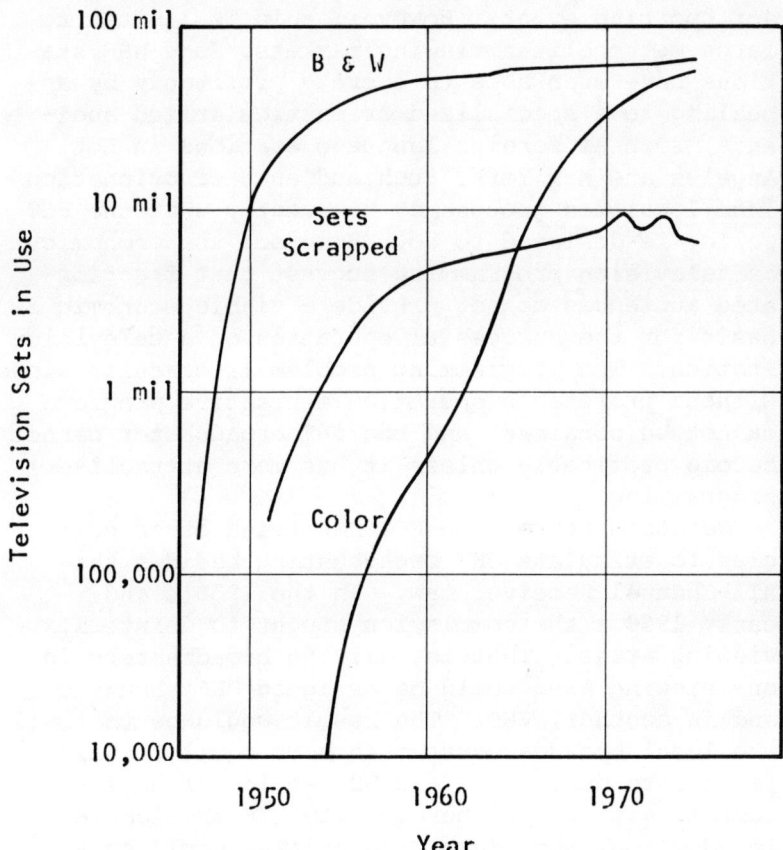

Figure 4-2   Television sets in use in the U.S., by year.
             Source: *Television Factbook*, 1974-75 edition.

Another possible reason for the UHF problem is
program content. Many UHF stations are not network
affiliates and rely on local programming. This

means that the programs available are movies, syn-
dicated programs, rerun serials, and sports events.
This type of program is unlikely to attract a sufficient audience to make the operation profitable.
(1) One exception to the lack of profitability of
local programming has been the televising of major sporting events. However, this is limited to
large metropolitan viewing markets. Some UHF stations have been able to operate profitably by ap-
pealing to a specialized or particularized audience, such as foreign language stations in Los
Angeles and New York. Such audience fractionation
into localized groups may be exactly what the FCC
policy is designed to do. However, the economics
of television programming suggest that fractionated audiences do not provide a viable economic
basis for the successful operation of a television
station. The programming problem is circular since
without profitable operation attractive programs
cannot be obtained, and the UHF broadcaster cannot
become profitable unless it has more attractive
programming.

Deintermixture. The FCC has tried other policies to stimulate UHF broadcasting besides the
All-Channel Receiver Law. In the 1950's and
early 1960's the commission sought to deintermix
viewing areas. That is, all the broadcasters in
one viewing area would be assigned UHF channels
and in another, VHF. The result would be that all
the local broadcasters would have equal technical
factors to deal with, and UHF would not have to
compete with the higher quality VHF broadcasters.
In addition, the viewers in an area could supposedly purchase either a UHF for a VHF television
set and be able to receive all the broadcasts in
the area. There was a good deal of political opposition to the deintermixture proposals, even
though the FCC did institute the policy on a trial
basis in several areas.(6) The VHF broadcasters
opposed having to give up their valuable and

qualitatively better channel assignments where
they faced a UHF assignment. The viewers in VHF
areas which were slated to become UHF areas seemed
to be opposed if for no other reason than that they
would have to purchase a UHF receiver. When the
FCC proposed the legislative action in 1962, it
combined certain deintermixture proposals with the
all-channel proposal. Then, to gain support for
the All-Channel Act among various groups, the commission agreed to drop the deintermixture proposal
in return for support for the All-Channel Act from
broadcasters and viewers.

Politics of the All-Channel Act. In order to
understand the politics of the All-Channel Act,
it is useful to reexamine the events leading up
to it. The FCC had early made a major decision
to foster diversity in TV programming through
licensing new broadcast stations in the UHF band.
Unfortunately, UHF reception is frequently of much
poorer quality and requires a different tuner than
VHF. Thus, few viewers were attracted in areas
where VHF channels were available, and few television sets were manufactured with capability of
receiving UHF channels. A UHF license, in many
areas of the country, was useless, even if the
UHF broadcaster was a network affiliate.

The FCC considered various policies designed to
increase the attractiveness of UHF channels to
broadcasters besides deintermixture. Along with
its effort at deintermixture, the FCC began to
lobby in Congress for the authority to require that
television sets be capable of receiving UHF. The
commission's proposal on deintermixture was opposed
by the broadcasters and the networks which stood to
lose VHF channels and the congressmen who represented the viewing areas that were designated for
deintermixture to all UHF. As a result the FCC
dropped its deintermixture efforts in return for
support from the broadcasters and networks for the
All-Channel Act. The equipment manufacturers gen-

erally supported the act even though it would increase the cost of their sets. The UHF broadcasters supported the act because they stood to
gain some viewers simply because the viewer could
now receive the UHF broadcasts. Some viewers apparently also supported the legislation because
they valued increased diversity. These people all
stood to gain something from the UHF requirement.
The people who would pay the costs of the act were
the new purchasers of television sets, but at that
time consumer interests were barely organized and
although several Congressmen mentioned the increased cost to the buyer (estimated at from $20
to $45 per set), no significant opposition came
from the purchasers of television sets.

The technopolitical model for UHF TV is shown
in Figure 4-3. The final outcome was the All-
Channel Act which was ultimately supported by the
established industry (networks and VHF TV broadcasters) and consumer groups (viewers against deintermixture) as an acceptable compromise. New
entrants (UHF TV broadcasters) also supported the
policy. The result of the decision was generation
of a substantial market for UHF tuners and an increase in potential UHF viewers.

CATV Systems

At about the time that the FCC was initiating its
UHF policy to achieve localism and program diversity among television broadcasters, cable television (CATV) had its beginnings. (Initially the
acronym CATV stood for community antenna system,
but is now universally understood to stand for
cable television.) CATV systems developed in rural
and mountainous areas where over-the-air reception
of television signals was difficult or impossible.
They operated with a high antenna, placed on a
mountain top, which permitted clear reception of
distant television signals. The received signals
were then amplified and distributed by a cable to

the homes of subscribers either a short distance
or many miles from the antenna. VHF reception, as
well as UHF reception, is hindered by distance and
other obstructions. Thus, various local entrepreneurs in these communities began to build cable
distribution systems to improve the television reception in the area.

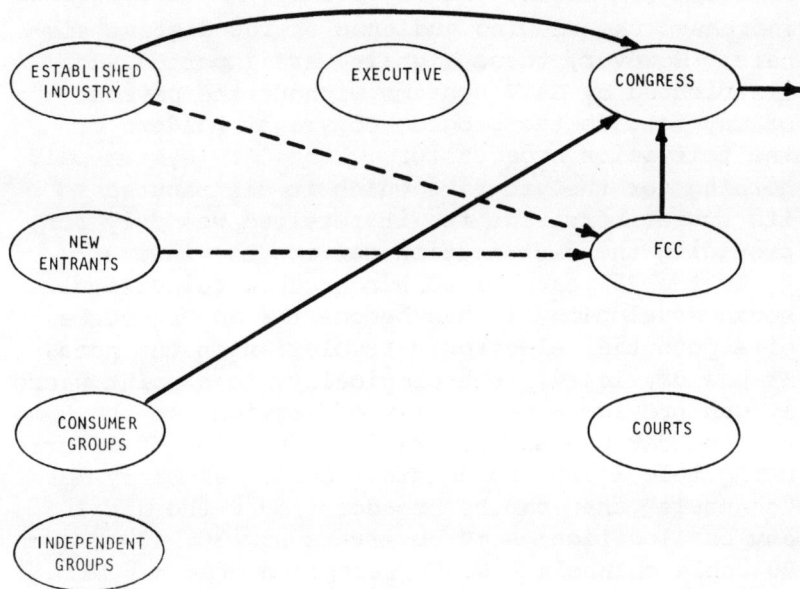

Figure 4-3 UHF TV All-Channel Act Model

Established industries included the UHF equipment manu-
    facturers, the UHF broadcasters and VHF broadcasters.
    The equipment manufacturers sought increased markets
    for UHF sets. UHF broadcasters stood to gain some
    viewers, and VHF broadcasters could defuse the dein-
    termixture issue by supporting the bill.
New entrants included potential UHF broadcasters who
    would enter the market if there were more potential
    viewers.
Consumer groups supported the increased diversity in pro-
    gramming.
The Federal Communications Commission also lobbied
    strongly for the enactment, to support its policy of
    localism in broadcasting.

One of the attractive features of CATV is that
it can receive distant signals by placing a pick-up
antenna near the TV station, relaying the signal
back to a central location, and distributing it to
the homes of viewers. For a monthly subscription
fee, the viewer can not only receive pictures from
local broadcasters but also can receive additional,
high-quality signals with the programming available
from large, distant viewing areas. In effect, this
increases the viewing audience of the distant signals. However, these signals were imported and
distributed by CATV systems without the payment
of any fees to the program copyright holders or
the television broadcasters. The CATV system paid
nothing for the programs which it distributed to
its subscribers. The fee it received was only for
providing the distribution service to viewers.

In the 25 years or so since cable television
began developing, it has become the major source
of a potential electronic revolution in the home.
It has developed, technologically, to a point where
it can provide a vast array of services to the home
ranging far beyond the original function of distributing television broadcasts. Cable can carry more
"channels" than can be broadcast over-the-air in
any one location; some observers say 30, 40 or even
80 cable channels. No TV reception area has more
than 12 to 15 operating stations.(2) Cable can also
provide a two-way capability so that subscribers
can respond to messages it carries. A cable system does not use radiated signals in the electromagnetic spectrum. While it may obtain a portion
of its programs from over-the-air transmissions,
the cable system, itself, does not broadcast.
Rather, it distributes all the signals through a
coaxial cable. The cable system conserves spectrum in that it does not utilize radiated signals
in its distribution process. While the distant
signals do use portions of the spectrum, no additional spectrum is required for their distribution
over the cable system.

The technical possibilities for CATV have led
some observers to call cable "the television of
abundance."(11) This technology presents the possibility of providing program diversity using many
distant signals to yield a variety of programs and
providing special purpose channels which, unlike
UHF, do not occupy any portion of the spectrum.
CATV systems in local areas are affecting the FCC
objective of developing UHF TV for local stations
and program diversity. As a source of program
diversity, cable does not suffer from the reduced
quality of pictures, which is a major problem with
UHF broadcasts. A CATV system can present a large
number of signals, all of high quality and with
considerable diversity. The distribution of UHF
signals over the cable can increase the potential
viewing audience for UHF.

The regulation of CATV has been done almost entirely by local governments. Through franchising
a CATV system, the local government gave one cable
company a monopoly in the viewing area, obtained
a franchise fee (usually a fixed percentage of the
subscription charge), and provided the local citizens with the opportunity to see programs not
available over the air. However, in the late
1950's, the established broadcasters began to agi-
tate for more extensive and protective government
regulation of CATV. Initially, the FCC declined
to regulate cable as a common carrier since the
CATV systems did not have the usual characteristics
of common carriers.(12) Then, under continued
broadcaster pressure, the commission sought statutory authority from Congress to regulate cable. In
the early 1960's Congress failed to enact such
legislation, and the FCC faced the need to regulate
cable without specific authority. Coming in the
back door, the FCC in 1962 began to regulate cable
systems which used microwave carriers to obtain
their signals.(13) The commission restricted the
kinds of television signals which the microwave

carriers could supply and thus the kinds of signals which CATV systems could obtain. In this
fashion, the FCC specified carriage and nonduplication requirements which limited CATV offerings.
These requirements were extended to all microwave-
served cables in 1965.(14) However, CATV systems
which did not use microwave as a source of distant
signals were unaffected by these FCC regulations.

During the 1960's the CATV systems expanded
greatly both in terms of the communities served
and in terms of the number of subscribers using
cable. Figure 4-4 illustrates the growth of CATV
systems and the increase of subscribers to those
systems. The data indicate that the major growth
in the number of subscribers began in the mid
1960's. Also, the number of subscribers per operating system began to increase markedly in the same
period. The saturation rate or number of CATV subscribers as a percentage of TV homes is 15 percent,
although in some markets, the ones in which CATV
has been operating for some time and where CATV
provides an attractive viewing alternative, the
penetration (homes served as percentage of homes
passed by cable) can range from 60 to 80 percent
(the national average is 55 percent). The growth
of CATV in the 1960's was not due directly to the
FCC regulations. In fact, the development was more
likely due to the affluence of the population in
the 1960's and the drastic increase in color television programming. The growth in subscribers per
CATV system indicates that the penetration rate
began increasing sharply in the mid 1960's.

In 1966, the FCC issued a Second Report and Order which exerted FCC jurisdiction over all CATV
systems and required that in order to import distant signals into the top 100 viewing markets, the
cable company had to have express commission approval.(15) The commission's concern with this set
of regulations was the fractionation of viewing
audiences in large-city UHF markets. The FCC saw

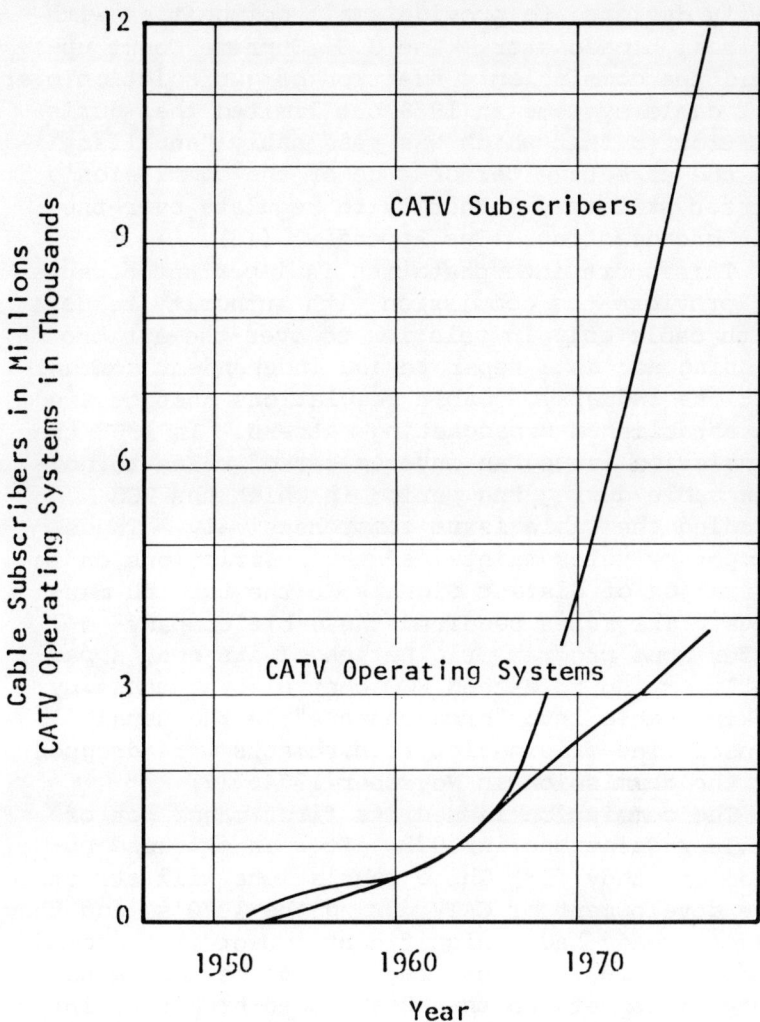

Figure 4-4   CATV systems in operation and CATV
             subscribers by year. Source: *Television
             Factbook,* Volume 46, 1977 edition.

cable as an important part of small community viewing, but the protection of its large-city UHF
broadcasting efforts was uppermost in the commission's mind, even though the UHF plan was origi-

nally designed to provide small communities with
a local broadcaster. The U.S. Supreme Court upheld the commission's exertion of jurisdiction over
all cable systems in 1968 but limited that jurisdiction to that which was reasonably "ancillary"
to the effective performance of the commission's
direct statutory authority to regulate over-the-
air broadcasting. (See Appendix).(16)

This court interpretation is important because
it provides the commission with authority to deal
with cable only in relation to over-the-air broadcasting not as a separate and independent communications industry. Cable regulations must be tied
to established broadcasting matters. In 1968 the
commission issued an interim set of rules to govern cable during the period in which the FCC
studied the cable issue comprehensively. These
temporary rules maintained the restrictions on importation of distant signals in the top 100 markets. The rules required the cable company to
offer some program origination of its own, apparently trying to extend FCC control over cable by
making cable into "broadcasters" in the usual
sense. The origination requirements were dropped
by the commission in November 1974.(17)

The commission issued its first major set of
cable regulations in 1972, after an extended period of study.(18) These regulations will structure
the development of CATV during the 1970's, and they
have already had a significant impact on the cable
industry.(19,20) These regulations reflect a partial
compromise between the established broadcast interests and the cable companies. The major gain for
cable was limited entry into the top 100 markets,
although there are complicated restrictions relating to signal importation and nonduplication of
local signals. When the regulations were issued,
the signal importation restrictions would allow all
of the top 100 markets to obtain at least one distant, independent signal over the cable and at most

three independents.(21) The effect of this was to
protect local network affiliates and to allow distant competition for local independents. The number of distant signals imported decreases as the
market position of the viewing area decreases.
Thus, localities that are smaller markets can import fewer signals.

As one of the major requirements of the 1972
regulations, the FCC stated that it was time for
cable to start providing some of the vast technological potential which many observers had discussed for sometime. To facilitate this, the FCC
requirements provided that all new (post-1972) and
all established cable systems (by March 31, 1977)
must have the capacity for 20 channels of offerings.(22) This compliance deadline has been extended until June 21, 1986.(23) In addition, the system
must be capable of providing rudimentary two-way
communication, so that the viewer can provide a
yes-no or on-off response to a transmitted request
from the head-end.(24) The regulations require that
the cable system provide one public access channel
each for such things as education, local government,
and a public forum.(25) However, these access channel requirements were reduced to one channel until
demand for more is evident.(26) These are to be provided free for the first five years of operation.
Table 4-2 indicates the recent development of channel capacity and two-way capability in CATV systems. The growth can only be shown for the years
since the 1972 regulations. However, the data indicate that an increasing proportion of systems
have the required capabilities. The extent to
which these capabilities are being used is unknown,
but the number of two-way operating systems is
quite small, and these are being tried on an experimental basis.

CATV and Copyright. A major problem which was
not settled by the 1972 regulations was in the
copyright dispute between cable companies and the

holders of program copyright privileges. Most of
the CATV industry were willing to pay some copyright fee for the use of programs. However, the
dispute was over how much the payment should be
and the means of making the payment. The Congress
repeatedly considered proposals to require such
copyright payments, because the Supreme Court
twice held that CATV systems do not "perform" a
program under the then existing copyright law and
therefore are not liable for copyright fees under
the 1909 statute.(28,29)

Table 4-2   Channel capacity of CATV systems

Year Capacity	Number of CATV Systems						
	1969	1970	1971	1972	1973	1974	1976
Not available	140	164	118	65	46	31	11
5 or less	572	520	421	387	336	267	189
6-12	1559	1720	1882	2026	2181	2320	2647
Over 12	29	86	157	361	-	-	-
13-20					262	293	424
Over 20					207	297	444
Total	2300	2490	2578	2839	3032	3190	3715

Source: *Television Factbook*, Vols. 39-46 (1969-77).

     In October of 1976, Congress adopted a major
revision of the Copyright Law.(30) It becomes effective on January 1, 1978, and one of its major
provisions is the inclusion "secondary transmission" by CATV systems under its provisions.(31)
Such cable systems must deposit royalty payments
and file accounts with the Register of Copyrights
for distribution to the copyright holders. The
act also creates a Copyright Royalty Tribunal,
empowered to determine the royalty rates for CATV
owners, among other things.(32) What effect these
regulations will have on the development and oper-

ation of CATV systems cannot be predicted at this
time. It is safe to say, however, that the copyright expenses which some cable operators will now
incur, will undoubtedly increase the costs of operation and make entry into the cable business less
attractive.

CATV Technopolitical Model. With the technology
of CATV, the opportunities for development are
present. The restrictive features of the situation depend on what is economically attractive
under current regulations. While economics can be
viewed as the primary controlling variable, the
economic market place is narrowly confined by government regulation and political pressure from
various interests. Thus, the primary factor is
more likely to be the openness with which the
regulator permits entry and utilization of the new
technologies.

Regulatory commissions have frequently been
charged with protection of the industry which is
supposedly regulated. Protection can come in the
form of excluding new, competitive entrants, or in
the form of rate protection from lower, competitive
suppliers of the service. As such a regulator, the
FCC has been caught between the legislative direction to regulate in the "public convenience, interest, and necessity," and the demands from established, politically-strong broadcast interests
which seek protection from various competitive
threats. The policy results of the FCC have often
been either in favor of the status quo--the established industry--or to permit only slight and
limited entry of new services or technologies. The
effect of such parameters on the development of
technology and its utilization would seem clear.
Generally, the new technology will not be permitted to compete with or replace the existing technology which has the advantage of being in-place
and operating. Thus, while cable technology provides a pandora's box of technical possibilities,

the realization of any of these will depend on
political decisions by Congress and the regulatory
agencies, as well as the general political climate,
the Executive Branch, and the courts. The current
cable regulations are a prime example of limited
competitive entry for the new technology in the
face of the established interest's pressure for
protection.

The major participants in the CATV area are the
established, over-the-air broadcasters, the networks, the program producers, the CATV operators,
and various equipment manufacturers (both for
cable technology and broadcasting). The results
of the political interplay of these forces vary,
but it appears that the established interests have
not been losers in terms of policy. The 1972 cable
regulations open the top 100 markets to CATV systems, but the restrictions on signal importation
and other programming limits indicate that the
over-the-air broadcasters in these markets are
still in a secure position. The cable interests
accepted the 1972 regulations as part of a compromise arrangement negotiated by the Office of Telecommunications Policy in the White House. Most
CATV operators saw getting a foot in the door of
the top viewing markets as a gain. However, as it
has turned out, the door was barely opened, and
current capital markets make it unlikely that many
CATV systems will venture great amounts of investment in these markets at this time.

The system model for the CATV allocation process
is shown in Figure 4-5. The ultimate policy represents a prevailing of the views of the established
industry (TV broadcasters), an independent group
(Sloan Commission), and the Executive Branch. The
new entrants (CATV companies) strongly opposed the
limitations on imported signals and the anti-
siphoning rules which adversely affected their
marketability. The net technological result was
a requirement to update existing systems to a

specified level but to provide little incentive for
further development.

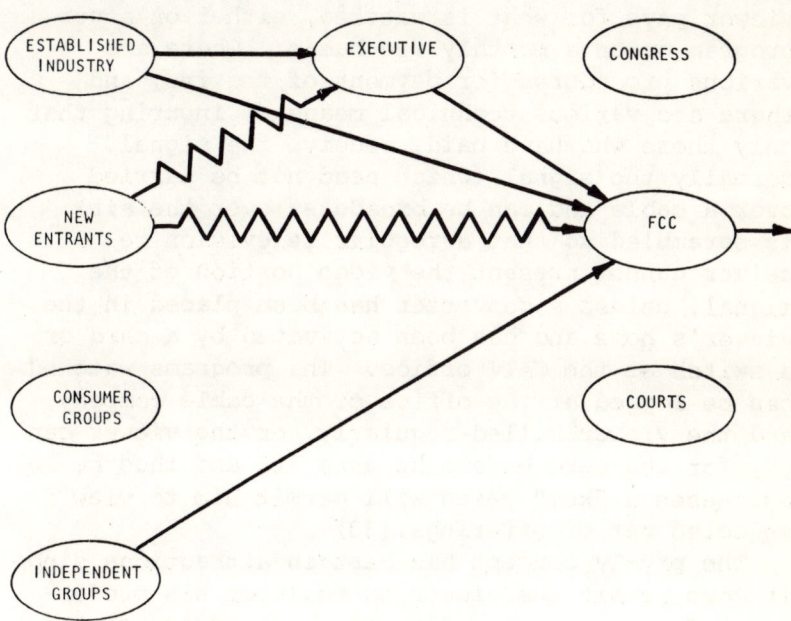

Figure 4-5 CATV allocation model

Established industries include operating, over-the-air
   broadcasters who sought to impose severe limitations
   on CATV operation in the top viewing markets through
   FCC regulations and White House pressure.

New entrants were the CATV systems, seeking to receive
   some economic and technical opportunities in the
   large viewing markets by means of favorable FCC
   rules.

Independent groups included the Sloan Commission and
   the Ford Foundation which provided studies of the
   capabilities of cable television systems and advocated a policy middle ground.

Pay Television

There are alternatives which CATV systems can pursue to obtain various kinds of income from viewers

other than providing distant, over-the-air signals.
One of the more promising seems to be some form of
pay-TV. Pay-TV is based on the principle that the
viewer pays for what is watched, either on a per
program or on a monthly fee basis. There are
various procedures for payment of the fee, and
there are various technical means of insuring that
only those who have paid, receive the signal.
Normally the signal (which need not be carried
over a cable and can be broadcast over the air)
is scrambled so that a regular television receiver cannot present the video portion of the
signal, unless a converter has been placed in the
viewer's home and has been activated by a card or
a switch at the CATV office. The programs watched
can be logged at the office of the cable company,
and the viewer billed regularly, or the viewer can
pay for the card before he uses it, and thus he
purchases a "key" which will permit him to view a
selected set of offerings.(33)

The pay-TV concept has certain attractions since
it does permit the viewer to register his preferences for programming directly by choosing to "buy"
or not to "buy" individual programs.(1) Thus, in
theory, programs which are not profitable can be
determined by the market immediately rather than
after viewer ratings have been registered, and advertising has sold or not sold particular products
to consumers. The individual program can be directly measured in terms of its economic power.

In addition to these aspects of pay-TV, the
broadcaster or CATV operator directly receives the
income from the programming he provides. That is,
the income is not obtained by the commercial advertiser who in turn pays the broadcaster for the continued presentation of the program. The cable
operator receives no revenue from the advertiser,
so the payment to CATV systems would be direct,
rather than indirect through subscriber fees. For
cable, the subscriber would still pay for being

wired into the CATV system and would then pay an
additional fee for the privilege of viewing the
selected programs on the pay-TV channel. A major
attraction of pay-TV is that it would permit a
cable system to present specialized programs, which
only a few viewers would watch and which would be
paid for by those who want to watch them. This
would have the effect of fractionating the audiences, but the viewer who wanted to see a mass
audience program or could not afford to pay for a
special program would not have to do that. Thus,
some see pay-TV as the means to "improving" the
quality of television programming and at the same
time permitting direct payment for those program
changes.

However, there are strong arguments against any
sort of pay-TV system, whether over-the-air or
cable. First, only those people who can afford to
pay for the service would be able to watch it.
While all viewers have to purchase a television
receiver, they do not have to purchase the advertised product on commercial broadcasts. Thus a
free-rider situation can and has developed in the
existing system since many people watch programs
but pay little if anything for those programs. It
appears that a pay-TV system does screen out particular, low economic strata of the viewing population. Either because they are unable to afford
the pay-TV service or because the programming is
not attractive to them they are screened out. The
experiments done with pay-TV did not attract many
viewers from these economic and social strata. The
extreme argument here indicates that eventually
the only television viewers would be those who
could afford it, and a large portion of the viewing audience would be eliminated. Furthermore,
the programming available under a pay-TV system,
if fully developed, would be much different than
the mass audience programs currently offered free
to the viewer. Whether this change in program con-

tent is desirable is debatable. Certainly some
view the "quality" of current programming as poor,
but "quality" depends on the tastes and preferences
of the individual and many see current programming
as acceptable in quality.

In addition to these economic factors relating
to pay-TV, the Federal Communications Commission
has strongly regulated the kinds of programs for
which a pay scheme can be arranged. "Siphoning"
of programs off the air or away from the commercial
broadcaster, as it is called, is viewed with great
suspicion by broadcasters and by the FCC. This
would allow the pay-TV proponent to purchase the
"cream" of programming and then charge for it.
The commercial broadcaster on the other hand, must
not only pay for the same program, without the
direct income for pay viewers, but must then set
a price to advertisers which is economically attractive before he can air the program. The result
of no anti-siphoning regulations would be, so the
argument goes, that the only programs the over-the-
air broadcaster could afford would be unattractive
to most viewers and thus the broadcasters' revenues
from commercial advertisers would decline to the
point where the broadcaster would be unable to make
a profit from his operation.

As a result, the FCC has imposed anti-siphoning
regulations which prohibit a large number of programs from being aired through a pay-TV system.
The regulations now permit showing only of movies
which are less than three years old or over ten
years old, and sports events which are not normally
aired live or have not been shown live during the
past five years by any local broadcaster.(34)

Thus, the only markets available to pay-TV are
current or very old movie markets and restricted
sports audiences. This may be an adequate market
for pay-TV to begin airing a channel of this sort.

However, the FCC has permitted over-the-air
pay-TV only on an experimental basis until quite

recently. The experiments indicate that the marketability of over-the-air pay-TV is primarily to
specialized audiences. Currently, some cable companies are providing pay-TV services either on a
regular basis in some selected viewing markets or
as a closed circuit pay-TV system for hotels. These
appear to be the primary points of entry for pay-
TV currently, and these are very limited efforts
at this time, although the Cox Cable Company has
25,000 pay-TV customers in three systems and Teleprompter has 32,000 in four systems.

Whether pay-TV develops will depend on several
factors, including the FCC siphoning regulations.
The economics of acquiring programs for pay-TV
audiences will have much to do with its success
or failure. The kinds of programs which cable
operators can provide for pay-TV viewers will depend on how many people are interested and how
much money they are willing to pay for the added
attraction of specialized programs, compared with
mass audience, commercial programming.

Radio Broadcasting

The radio broadcasting industry is the oldest user
of the electromagnetic spectrum to reach mass audiences in the United States. The technology was
originally developed in the years before and during World War I, and the first extensive commercial
use occurred in the 1920's. The original technology involved amplitude modulation (AM), and only
later, in the 1930's, was frequency modulation (FM)
developed to provide a higher quality signal than
AM. The AM system of broadcasting is the most
heavily used in this country, in part, because it
was developed earlier, in part because it does not
use as much spectrum, and in part because its lower
transmission frequency provides much greater geographical coverage. A frequency allocation for AM
broadcasting involves 10 KHz of bandwidth while a
single FM broadcast frequency requires 200 KHz of

bandwidth. Thus, approximately 20 AM radio broadcasters can be accommodated in the bandwidth required for one FM broadcaster. In addition to the
historical and technological reasons for the
heavier use of AM broadcasting, the FCC's regulatory response to these two kinds of broadcast
techniques has been different in the early years
of each one's development.

Figure 4-6 provides estimates, by year, of the
number of AM and FM broadcast stations on the air.

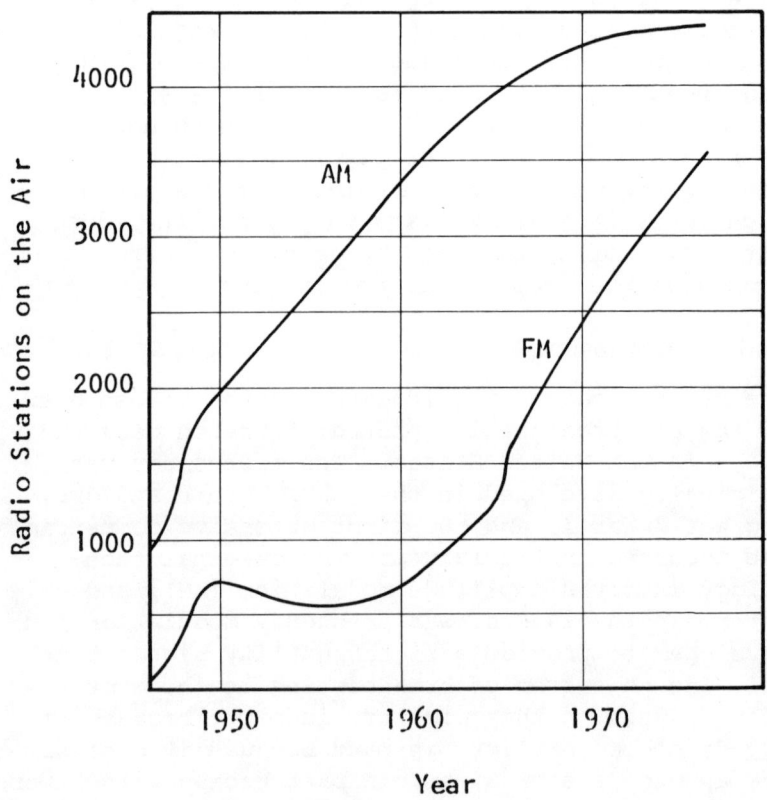

Figure 4-6   Radio stations on the air in U.S., by year.
             Source: *Television Factbook*, Volume 46,
             1977 edition.

These data show several things. First, even
though data for the first 20 years of AM broadcasting are not presented, it is clear that the
number of AM stations has grown primarily in the
period from 1946 (913) to 1974 (4407). Thus, the
intensive development of the AM spectrum has occurred largely in the past 30 years and is probably due largely to the technological developments
of this period. Such devices as directional antennas and limited FCC license allocations (limited to specific times of day or power) have permitted a tightly packed allocation for AM.(2)

A second general conclusion regarding these data
is that the development of FM, occurring wholly
since World War II, has been much more gradual than
that of AM until recently. One of the major events
was a reassignment in 1945 of FM to a higher portion of the spectrum. This came in connection with
the initial allocation of VHF television channels
and the reassignment resulted from the desire of
VHF television broadcasters to have the initial
assignments which had been made to FM on an experimental basis in the early 1940's.(6) One result of
this reassignment was that the original FM receivers and transmission equipment were made ineffective. The FM broadcast industry had to begin
afresh, with new transmitters and new receivers
after the reassignment. In addition, the cost of
manufacturing FM equipment has been relatively high
until quite recently. While an AM receiver costs
varying amounts from less than $20 upward, the cost
of FM receiving equipment has been much higher.
Only within the past five years or so has the cost
been reduced to the point that people could buy
such equipment easily and at a reasonable price.
The technology has only recently reduced the cost
of such equipment to within the reach of most radio
users.

Figure 4-7 shows the estimated number of radio
receivers in use from 1950 to 1974. The growth

rate of radios in use has been much greater than
the growth rate of population in the United States.

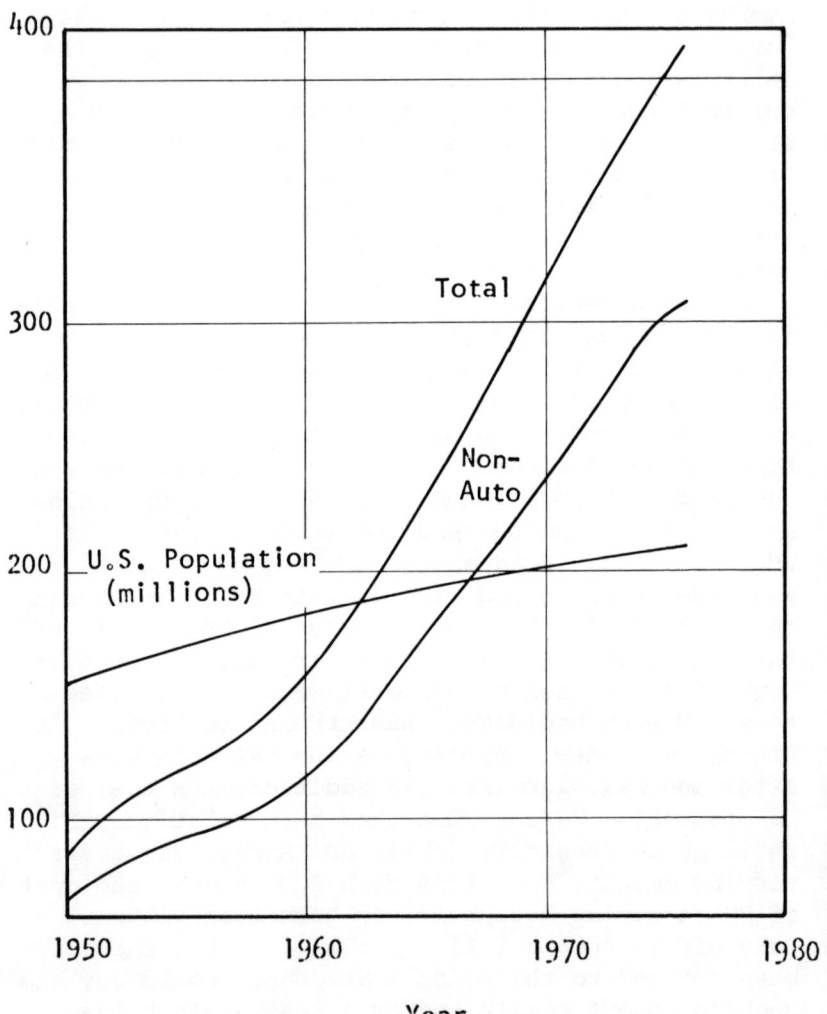

Figure 4-7   Estimated radio sets in use in U.S., by
              year. Source: *Television Factbook,*
              Volume 46, 1977 edition.

In fact, the number of receivers exceeded the population in 1963; this event occurred at about the
time that the transistor became widely used in the
construction of inexpensive radio sets. Since that
time, the number of sets and the rate of increase
has been much greater than the population, probably
due to further improvements in the technological
capability to build inexpensive receivers. The
number of receivers in automobiles has grown at a
much slower rate than "other" receivers. This is
important to note in light of recent FCC efforts
to initiate automobile radio receiver policies.
This subject is discussed below.

Most of the growth of FM broadcasting has occurred since the early 1960's.(35) The assignment
of construction permits to FM stations has exceeded the number on the air, just as the AM
assignments are greater than those on the air. However, the proportion of AM station assignments
that are on the air has consistently been much
higher than the proportion of FM stations on the
air. Table 4-3 shows the percentage of stations
on the air of the total licensed. Although FM
percentages have been steadily increasing, they
have been only rarely above 90 percent while AM
has been above 90 percent since 1950. This might
suggest something of the financial difficulties
FM broadcasters have in getting the needed capital,
in comparison with AM broadcasters, to build and
begin operation of the station. In addition, it
may also suggest the relatively better chance of
obtaining FM licenses than AM licenses in most
areas.

The pre-engineered allocation table used for
assignment of FM broadcast frequencies has not
produced the same kind of congestion incentives
for technological innovation which AM has seen.(2)
Since the bandwidth of FM stations is known, and
each FM frequency has been designated and is waiting for a potential broadcaster, a new entrant

knows what frequencies are still available in his
broadcast area, when he seeks a license. Thus,
the "protected" status of FM frequencies means that
there is not likely to be much outside pressure unless there are more applicants seeking licenses in
a region than there are allocated stations. So far
that has not occurred, even in the largest population centers.

Table 4-3   Percentage of AM and FM licensed broadcast
            stations on the air, 1946-1974

Year	AM	FM	YEAR	AM	FM
1946	87	16	1960	92	80
47	65	20	61	96	80
48	81	37	62	94	79
49	88	70	63	96	80
1950	91	91	64	94	83
51	93	95	65	97	75
52	95	97	66	98	88
53	94	94	67	98	86
54	93	91	68	98	88
55	96	94	69	99	91
56	96	96	1970	98	92
57	96	94	71	99	90
58	97	91	72	97	91
59	96	83	73	98	90
			74	98	89
			75	98	89
			76	98	90

Source: *Television Factbook*, Vol. 46 (1977).

FM programming has developed in a specialized fashion. Based largely on the high fidelity of FM broadcasting, the programs have frequently been designed to appeal to audiences who are interested in high quality sound reproduction, such as opera or classical music enthusiasts. More contemporary programming has appeared on some FM stations, but some of the highest revenue producing programs, such as those aimed at teenage groups, are often not carried on FM. Even if these were broadcast on FM, the absence of a large supply of FM portable and automobile receivers, would mean that the broadcaster would reach only a small portion of the potential market.

In recent years audiences have become more concerned with the quality of reception. Frequently such individuals have their own custom sound systems as an alternative to listening to FM radio. Stereo and quadraphonic sound systems tend to replace FM broadcasts among many potential FM listeners. The alternatives to FM listening are likely to be more attractive because they permit listeners to select exactly the kind of music or other entertainment they desire. At the same time AM radio provides an acceptable level of quality for a large segment of the listening audience.

The Federal Communications Commission has pursued a variety of development policies for FM. Congress has considered proposals which would require that all automobile radios have the capability of receiving FM broadcasts.(36) This is patterned after the UHF All-Channel Receiver Act and is designed to create a larger FM audience. It would certainly create a market for more FM receivers. The FCC has found that there is a "correlation between FM earning ability and the sale of FM automobile radios," since the prime advertising time on radio is the time of day that many people are in their automobiles going to or from work.(37) Interestingly, the same efforts at

development are being considered for FM as were
tried for UHF television. Current organized consumer opposition to this policy indicates that
potential political interest groups change over
time and that what was achievable once may not be
possible at another time.

FM Automobile Receivers Model. The interplay of
the political forces involved in the FM automobile
receiver decision is shown in Figure 4-8. Although
the final decision on the requirement for FM in
automobile radios has not been made, the strong
opposition from a consumer group (labor union)
appears to have had a significant impact on the
outcome. This opposition has been effective in
spite of the efforts by the FCC and established
industry (set manufacturers and FM broadcasters).
This example illustrates the potential effectiveness of organized consumer input to the allocation
process and how it can affect the directions of
technological development.

AM Broadcasting. The development of AM broadcasting technology has been much different from
that of FM. The AM spectrum came into use when
radio broadcasting first developed in the 1920's,
so it has had a longer history.

It is evident from Figure 4-6 that there are
more AM stations on the air than FM. In 1939 the
AM spectrum was supposedly "saturated" and it was
believed that it could absorb no more stations.
Earlier, the Federal Radio Commission (the prede-
cessor to the FCC) had taken 150 stations off the
air in an effort to reduce the congestion.(2) However, since 1939, in exactly the same amount of
spectrum bandwidth (1 MHz), the number of AM stations on the air has grown to six or eight times
the 1939 number. There is no pre-engineered allocation table for AM. This means that the number
of broadcasters is not fixed for a region ahead
of time. Rather, depending on time of broadcast,
antenna direction, and power, the same spectrum can

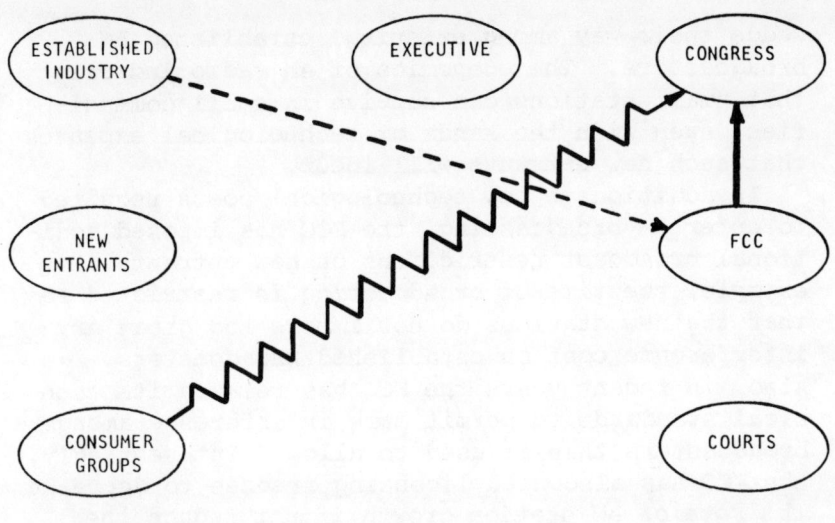

Figure 4-8   FM automobile receivers model

Established industries include FM equipment manufac-
    turers and FM broadcasters, both of whom would gain
    from FM receiver requirements.

Consumer groups include both purchasers of automobiles
    and labor unions who oppose the imposition of costs
    of the receiver on the purchaser of automobiles.

The Federal Communications Commission is strongly
    advocating the adoption of the policy.

be occupied by a large, and unspecified number of
stations. Licensing has been on an ad hoc basis.
The congestion which this produces has given broadcasters incentives to develop new technologies to
crowd more broadcasters into the same spectrum. (2)
The overcrowding has largely been solved by the
use of more sophisticated and more expensive directional antennas. This places the cost of technological development on the newcomers who seek to

wedge their way among existing, established AM
broadcasters. The economics of AM radio indicate
that small stations can survive in small communities, even with the kinds of technological expenses
that such new entrants will incur.
     In addition to new technological costs required
to enter AM broadcasting, the FCC has imposed additional broadcast restrictions on new entrants. For
example, the time of broadcasting is restricted so
that the new stations do not impose too great an
interference cost on established broadcasters.
Also, in recent years the FCC has relaxed its technical standards to permit more interference among
broadcasters than it used to allow. Interestingly,
the FCC has also used licensing freezes to decrease
the rate of AM station growth if not reduce the
number of AM broadcasters on the air. The most
recent freeze on new AM licenses was from July
1968 to April 1973.(37) The period of nearly five
years did show a decrease in the growth of AM stations on the air. See Figure 4-6. The freeze did
not, however, reduce the number of stations on the
air.
     Since the beginning, AM broadcast policy and de-
velopment policy has been entirely ad hoc. Based
on AM technology, the broadcasting industry gave
rise to the need for regulatory intervention in
the allocation of spectrum. These broadcasters
preceded the FCC and for 50 years have been the
established, broadcast industry. The lack of systematic policy in this area may be disconcerting
to some, but the technological growth in AM suggests that for some kinds of development a conges-
ted spectrum may not be a significant deterrent.
Without a systematic set of policies by the regulatory agency, the AM spectrum has become heavily
developed for commercial broadcasting. The AM
spectrum has absorbed an increasingly larger number of broadcasters and has continued to attract
audiences through low receiver costs and by specialized programming.

AM programming is specialized, but it lacks the
high quality of FM reception. The programming tends
to attract larger segments of the mobile population. Thus, popular music stations are numerous
because of the potential market this kind of programming reaches. In some larger population centers, there are all-news and information stations,
and in rural areas the stations can specialize in
country and western music and programming. Not
all segments of the listening population are satisfied by current AM programming. However, the segments of the audience which support such stations
economically are provided with the programming
desired. Part of the audience attraction for AM
radio involves the inexpensive, easily-transport-
able radio. People who carry a receiver around
with them, on their person, or in their automobile,
provide a large part of the radio audience. This
audience phenomenon might be called the "pocket-
radio syndrome," but, by whatever name, the AM
broadcasting industry has become economically
viable on the basis of such audiences.

In addition to these considerations, it can be
pointed out that the AM broadcast industry is a
prime example of local programming and diversity.
It is easy and economically feasible for many small
communities to support such a radio station. This
FCC objective for broadcasting may not be viable
for the television industry yet, but, in terms of
AM broadcasting, local diversity has been achieved.

The participants in the policy processes relating to radio broadcasting are similar in kind to
those in television broadcasting. The broadcasters
themselves are a central and potent interest. The
new broadcasting entrant can be viewed as an out-
sider, who must dislodge the established operator
to begin operation. However, the difficulty of
dislodging the status quo depends on the portion
of the spectrum involved and the amount of bandwidth which the new entrant requires. Thus, there

is room for more FM stations than currently are on
the air, although a particular applicant may face
an allocation plan that does not allow his station
since that broadcast region is full. Despite the
growth of AM stations, the only likely limitation
on new entrants now would be another FCC licensing
freeze or the lack of viable economic basis for
the operation of the new station.

Equipment manufacturers are not directly involved in policymaking, although they may be major
beneficiaries of a policy. This is especially the
case when the equipment requires a new device
(such as automobile FM receivers) which the pur-
chaser must buy from the manufacturer. The consumer or listener groups may be emerging as more
important and potent political, as well as economic, forces on current issues involving the
imposition of direct costs for equipment or services.

The important policy parameters in radio broadcasting appear, in part, to be the level of signal
quality possible and desired by the audience and
the presence or absence of a pre-engineered allocation table. Had the FCC set the maximum number
of AM broadcasting stations per listening area
beforehand, and had it designated frequencies for
them, the growth of AM would have been much different than it has been. While the resulting congestion would have been much less, the development
of AM technology would have been much different.
Whether this method (the allocation table) can or
should be used to achieve technological development
is an open question, but it appears that it can
achieve some technology development by creating
the incentive for developers to invest the time
and other resources necessary to produce new technological devices.

Conclusions about
Broadcast Technology and Regulation

The use of a technological device will depend on
factors such as economic costs of the device and
the benefits which it produces for the users. The
regulatory climate in which the technology is set
is as important as those economic factors. Certainly, there will be no demand for a device unless
the regulator permits the use of it. Furthermore,
the kinds of use permitted by the regulator are
crucial to the development of the technology just
as the permissable uses structure the economic
viability of various industries built around the
technology.

The regulatory parameters in which broadcast
technology exists illustrate some of these points.
For example, the degree and the precision of allocating a frequency may profoundly affect the development of a technology. If the allocation
involves predetermined frequency designations as
for TV and FM broadcast stations, there may be
little need for intensive development of the spectrum arising from "congestion" in the spectrum.
Another example of this is the choice of the color
television broadcasting system for the United
States. The FCC eventually chose the one which was
compatible with the existing, 6 MHz bandwidth black
and white system. There were several political
reasons for the FCC choice--such as not to make
obsolete the then existing television sets and not
to bifurcate the audience into color and black-and-
white viewers. A better quality color picture system was rejected, and the color technology developed
around the alternate technological choice.

The present allocation plan for UHF television
broadcast channels illustrates the protection of
an industry and the resulting absence of technological incentives for alternative utilization of
that portion of the spectrum. Under the FCC plan

1,098 assignments are possible. They are rather
loosely packed within the allocated, 70-channel
spectrum. Within the same 70 channels it is possible to make from 3,850 to 8,800 UHF channel
assignments and still meet the current FCC technical standards for UHF broadcasting. These
numbers may be surprising, but in comparison with
the current 1,098 UHF assignments that have been
designated, the actual number of UHF broadcasters
on the air are even more surprising. The data in
Figure 4-1 are only part of the numbers. There
are 184 commercial UHF broadcasters and 149 educational UHF broadcasters in operation, a total of
333 in comparison with the 1,098 FCC (30 percent)
designations and the technically possible 3,850
(8 percent) or 8,800 (3 percent) channel assignments.

The efforts to create artificial markets for
UHF broadcasters by requiring UHF tuner capabilities and limiting cable television in various UHF
viewing markets have not resulted in the policy
objectives of local broadcasters, diverse programming, and "full" use of the designated portions of
the spectrum. In contrast, the demands by alternative users of the UHF spectrum have mounted in
recent years. The UHF technology has neither had
to develop in the face of increased congestion nor
has the UHF broadcasting industry developed into a
viable enterprise.

In contrast to this, the AM radio spectrum has
accommodated increasing numbers of broadcasters,
and these in turn have served increasing numbers
of receivers. The spectrum congestion which resulted from the regulator's ad hoc approach to
license assignments has produced a variety of technological devices to permit more broadcasters to
use the spectrum. A number of small, local AM
broadcasters are able to provide adequate local
advertising and programming and are economically
viable.

The regulatory arena which structures the development of broadcasting technology is largely
defined by the participants who are involved in the
regulatory and the allocation process. The allocation of the scarce resource of spectrum is the heart
of the political process which determines the development and use of these technologies. The Federal
Communications Commission, a known arena in which
the participants can interact, handles this process
which largely revolves around who is assessed the
costs of the allocations and the acceptance of
these costs and benefits by the participants who
benefit from spectrum usage. The efforts to create
demand for UHF TV spectrum allocations among potential broadcasters through requiring receiver
capability illustrate the most open effort at technological incentives through allocation. To date,
these efforts have not achieved the desired results
from the regulator's or policy maker's perspectives. However, certain groups have benefited,
such as the UHF tuner industry. The failure to
enact a statute requiring FM tuners in automobiles
may simply be explained by potent, counter pressure
from consumer groups who would have to pay the
costs. Such pressure was not present when the UHF
All-Channel Act was proposed and enacted over a
decade earlier.

The technology associated with CATV systems
presents a picture of technology waiting for the
appropriate political and regulatory climate. The
technology has been available for some years. It
could develop rapidly and dramatically from its
present state but will not unless there are economic incentives for that growth. The economic
demand for advanced cable technology depends on
the regulatory policies of the FCC. Such demand
also depends on entry into the large markets, and
that entry is controlled by the established broadcasters and the commission. Until the broadcasting
interests, with program producers, are willing to

accept significant entry into these viewing markets by CATV systems, cable technology will develop
very little. The current FCC regulations which require certain channel capacities and two-way capability, are little more than window dressing in
terms of technology. The economic constraints on
cable development will be controlling as long as
the regulatory agency and established interests
impose markets constraints on CATV.

The entire regulatory scheme in this industry
is premised on a variety of imperfections in the
operation of the free market system which econo-
mists and entrepreneurs use for analysis. Regulation is an effort to interject certain controls to
prevent the development of undesired economic
effects which are associated with monopolies. One
of the current problems with regulation of industry
by the government is that the regulation may produce a variety of undesirable side effects of its
own. One of those is alleged to be protection of
the regulated industry from competitive entrants.
In the case of CATV, this protection of the established broadcasters seems to be quite evident in
the 1972 CATV regulations. The UHF allocation
system shows the effects of closing off a portion
of the spectrum to certain alternative uses in an
effort to encourage other, apparently more desired,
uses. The AM broadcasting industry, however, illustrates the technological results when some
element of open entry is permitted. That result
may or may not be desirable. However, this setting does have different technological results
from the closed-entry, regulatory effort.

Chapter 5

Two-Way Telecommunications

Importance of Two-Way Telecommunications

Just as radio and television strongly affected the social structure of the United States so also did the telephone. Radio and TV introduced national product advertising, new entertainment sources, and new means for influencing public opinion. The telephone produced an enormous impact by radically altering the methods and greatly increasing the speed of carrying out commercial transactions and in changing the nature and scale of direct inter-personal telecommunication. All of the industrially developed countries of the world felt the impact of the telephone.

Other means of two-way communication, besides the conventional telephone, are also vital to today's world. Two-way radio, one of the most important, is used for ship-to-shore communication, police and emergency communication, aircraft communication and control, dispatching of commercial vehicles such as taxicabs and trucks, and a wide range of military applications. The development of radio prior to World War I completely altered naval tactics when it became possible for ships to communicate with each other in darkness and over the horizon. This revolution in tactics has progressed until now virtually all military operations are based on continuous two-way communications with command centers.

In recent years a new type of communication has become increasingly important. This is the com-

munication of data as opposed to verbal information. The accumulation and analysis of data have
always been of great importance in operating government and commercial enterprises. However, with
the recent advent of the computer as a vital link
in the day-to-day operation of organizations and
businesses, the need for more rapid interchange of
data has greatly increased. Today, vast networks
are planned specifically to transmit information
to and from computers, and the volume of such communications will increase for many years to come.

Over the years introduction of new technology
to expand or improve two-way communication capabilities has continued. Perhaps the most dramatic,
recent technological innovations have been the communication satellites and the video telephone both
of which truly represent the integrated circuit
society. Of the two, the communication satellite
has been an unqualified success while the other
did not even get so far as a full-scale trial.
Later sections of this chapter explore the reasons
for this dichotomy of results.

Other less dramatic, though equally important,
developments have also been evolving in the past
two decades. Among them are the introduction of
new switching systems, the development of new data
transmission techniques and systems, the opening up
of the switched telephone system for attachment of
devices other than telephones, and the provision of
a variety of new services such as dial up remote
terminal connection to centrally-located general-
purpose digital computers. Enormous progress has
been made in many areas of two-way telecommunications and these developments have had a profound
impact on society in all aspects of its structure
and operation.

While these revolutionary developments have been
occurring in many areas of telecommunications technology, other areas have not undergone such development. Frequently, already available technology

has not been incorporated into existing systems in a timely manner. Why do we not have an effective mobile telephone capability? Or why is there great disparity in the effectiveness of the switching system technology used in various countries of the world and even within individual countries? Why has citizen's band radio become the most rapidly growing radio service in the United States but appears to be operating without enforcement of the regulations established to prevent its abuse and its interference with other services?(1) These and other questions concerning the forces that determine the support for development and adoption of new technology are examined in the following sections of this chapter.

Growth of Two-Way Telecommunications

Before considering case studies relating to specific examples of development and usage of two-way communications technology, perspective can be gained by considering the rate of growth and the scale of present utilization of two-way communication media. A convenient vehicle for this purpose is the switched telephone system, both in the United States and in other countries. Figure 5-1 shows the growth of various telephone services in the United States over the past two decades. By 1974 the telephone system in the United States represented an investment in excess of $80 billion and was still growing at a substantial rate of more than $8 billion per year. While services were expanding, the real costs of using those services were going down as the economies of scale became effective. Figure 5-2 shows the charges for long distance calls from New York to Tokyo, London, San Francisco, and Philadelphia. The costs have been referred to the 1974 dollar by dividing the actual charges by the ratio of the commodity price index for each year to that of 1974. It is evident from this figure that the greater the distance between cities called, the greater the reduction in costs that has occurred.

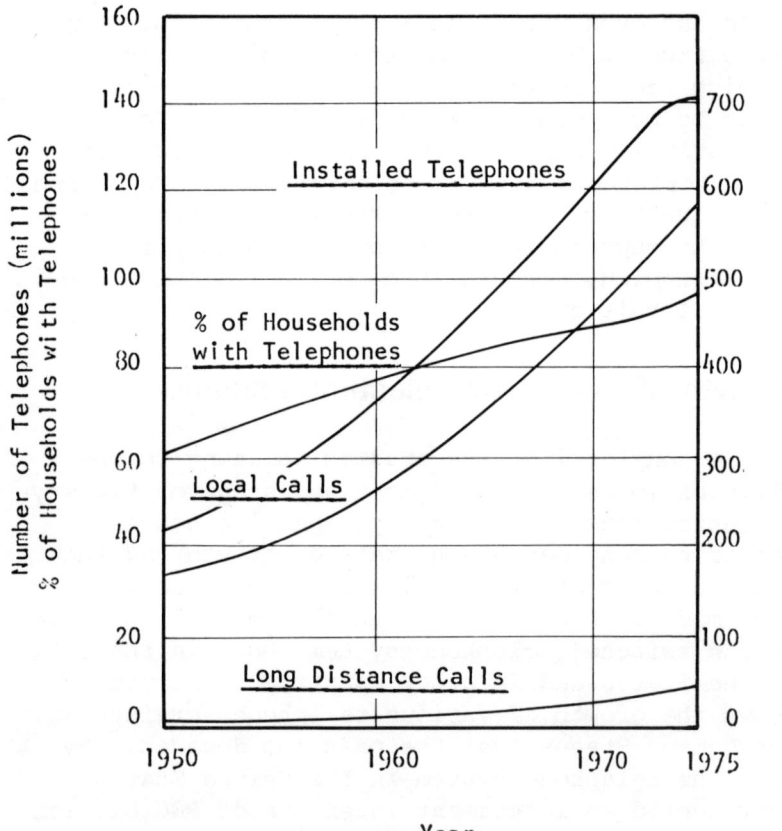

Figure 5-1 Growth of telephone services in the United States. Source: *Statistical Abstract of U.S. 1976*.

Elsewhere in the world, telephone installation and utilization has also increased. Figure 5-3 shows the number of telephones installed per 1000 population for the eight largest telephone users.

Clearly, there is a continuing upswing in telephone availability throughout the industrial nations. In none of the large countries does any
saturation of the market appear. Many cities, such
as Washington, D.C., actually have more telephones
than people.

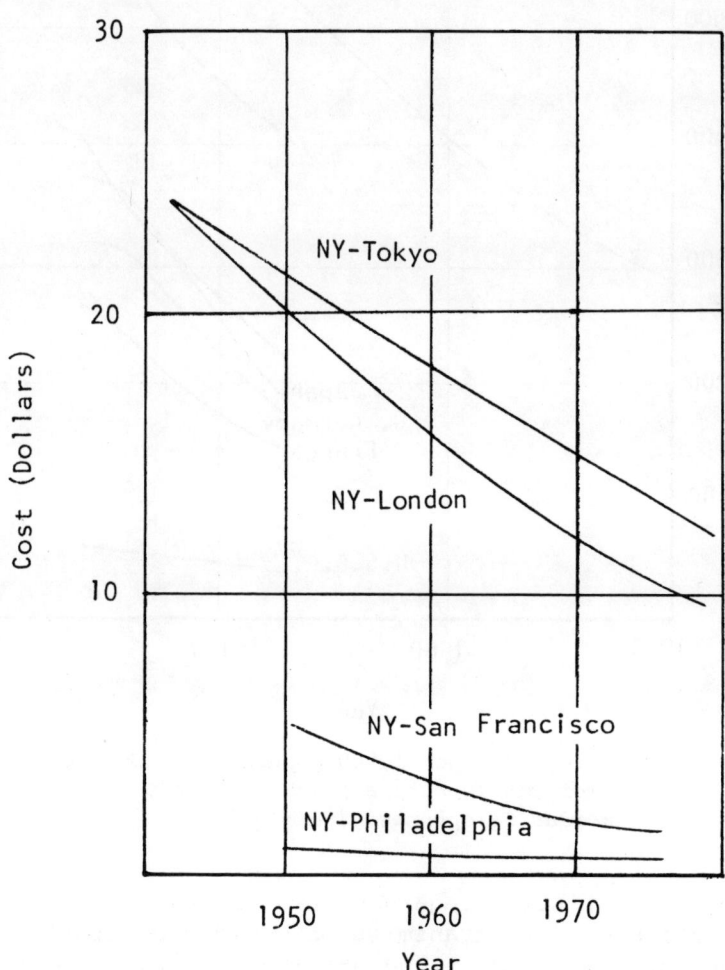

Figure 5-2   Long distance telephone charges adjusted
             to 1974 constant dollars. Source: *Statis-
             tical Abstract of U.S. 1976.*

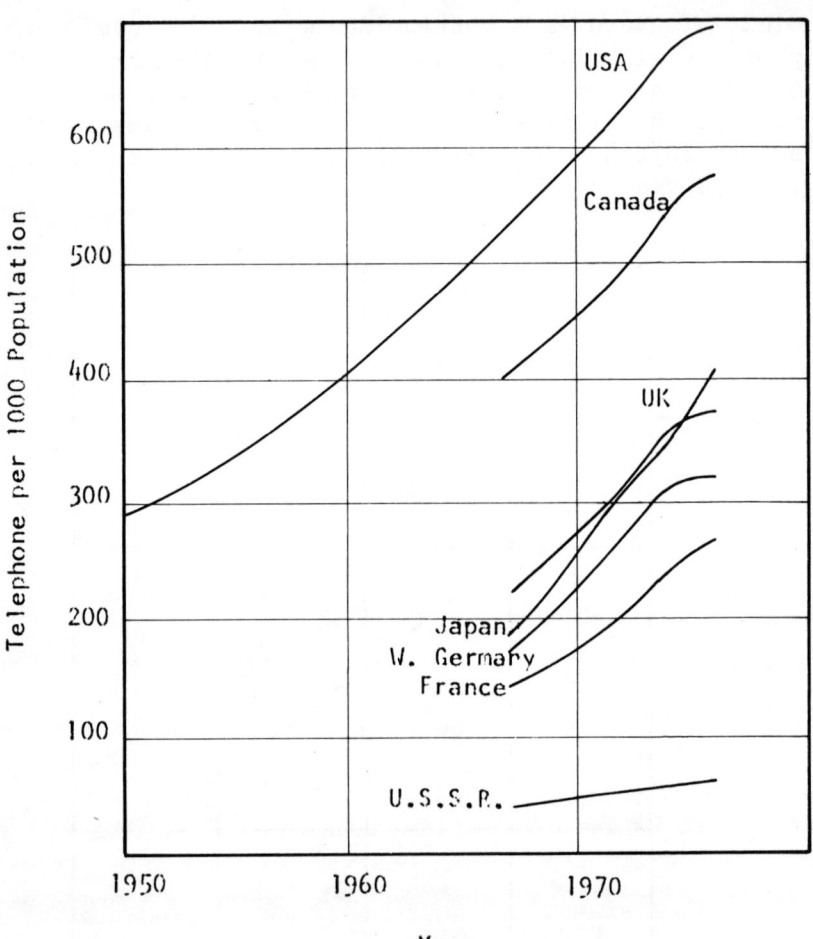

Figure 5-3 Telephones per 1,000 population. Source:
            *American Telephone and Telegraph Co.: The
            Worlds Telephones, 1961-1971.*

    Other statistics regarding two-way radio com-
munication or data transmission would show similar
strong growth trends. Undoubtedly, the availability and usage of two-way telecommunications is becoming more and more pervasive in world societies.

The effects of this phenomenon will probably be
felt for generations to come and could well alter
future world social, economic, and political structures. If so, then it is important to inquire into
how the rapid evolution of technological services
and methodologies is being controlled and directed
toward some set of socially acceptable and desirable goals.

A first reaction to looking at the advances in
telecommunication technology is to marvel at the
revolutionary developments that are occurring.
Many people take it for granted that such advances
are the natural outgrowth of well-ordered industrial research programs and that they find their
way into usage under control of social and economic forces that tend, in general, toward promoting the "common good." Unfortunately, this view
is far from the truth and completely ignores some
of the most important forces that are shaping the
technological, economic, and social character of
telecommunications developments.

Conventional economic factors such as basic scientific and engineering knowledge, available capital, and customer demand are necessary for successful development and marketing of new products or
services. However, particularly in two-way telecommunications, political decisions play an equally
and oftentimes greater role in determining the
course of technological developments. The political climate and the political arena of society
strongly influence how, when, and, frequently, in
what directions new technological developments will
occur. Often there is little relation between what
is technologically feasible and what the political
climate will permit. This aspect of technological
development will be detailed in specific case
studies and in the following discussions of the
regulatory process by which telecommunications are
controlled in the United States. The complex interplay of the various forces affecting telecom-

munications development will be clear. A subsequent chapter will consider alternative means of
controlling the result of these forces as a way
of approaching a regulatory process serving the
"common interest."

The Regulatory Setting

Common Carrier Political Arena

Regulation of common carriers is closely related
to regulation of the radio spectrum, except that
the participants and the issues differ. The Federal Communications Commission is the principal
arena in which the decisions are made, although
some participants in certain cases have used the
courts and the legislature. The actors include
the established common carriers such as AT&T and
the other telephone companies. Any new entrants
are either equipment manufacturers (other than
Western Electric, AT&T's equipment affiliate) or
small companies that propose to offer a service
for which they seek FCC approval. The new entrants are often specialized carriers, who seek
to provide a particular service to all comers.
They may be established firms which are expanding
or branching off into new areas, or they are new
companies just starting up with the proposed service as their only product. The service users, the
customers of the common carriers, seem to be more
active in the regulatory process than in the broadcast area. (Two-way communication does require a
less passive role for the users.) At least some
users, and nearly everyone in the nation uses the
telephone system to some extent, have been quite
vociferous, and sometimes successful, in achieving
a policy result beneficial to themselves.

Natural Monopoly

The telecommunications industry has been regulated
by various agencies of the federal government and

some state governments for most of the 20th Century. (2) Interstate telephone rate regulations began in 1910 when Congress enacted a statute giving the Interstate Commerce Commission authority to approve such rates. The regulatory arena shifted to the FCC after the enactment of the Communications Act of 1934. (3) However, the basic reason for regulation, market failure, has not changed much. That is, since the market pricing system does not provide an optimal distribution of goods and services, government has made an extensive effort to inject economic constraints to produce results comparable, in some respect, to a market situation. The natural monopoly, the *raison d'etre* for most regulation, is based on the economic fact that in some industries economies of scale permit a single supplier to provide complete service at the lowest price to customers. (4) In such a "monopoly" situation, the regulator is charged with protecting the "public interest" by preventing monopoly profits yet allowing the monopoly to provide its services at a profit. A variety of economic problems surround the existence and even the definition of a natural monopoly, and these problems must be addressed by the agency in order to achieve regulation in the "public interest." (5) The two-way communication system in this country has developed as a regulated industry. Federal, state, and local government agencies all have received regulatory authority on the basis of natural monopoly arguments.

The regulation of the telecommunication industry in this area is largely rate regulation rather than service regulation. The common carrier concept is based on a natural monopoly which offers service to all customers without regard to content, so long as the customer pays the allowable rate. It depends largely on regulatory control of the rate of return which these carriers can make on their capital investment. Rate-of-return regulation involves com-

plex considerations relating to what capital items
the carrier should be allowed to include in its
base--the rate base upon which the rate of return
is calculated. Within the past 10 years the FCC
has begun to re-examine its basic guiding philosophy regarding the common carrier telephone companies in the country and to pursue policies designed to achieve regulatory ends, by means not
generally accepted among the regulated industries.

Competition

Case studies in this chapter offer some recent examples of regulation in the telephone industry,
which contain an element of competition. The FCC
has made several decisions regarding competitive
entry into certain telephone services, which indicate that the commission may be trying to inject
some competition into the two-way communication
market for a variety of unarticulated or unclear
reasons. The emphasis of these decisions has been
to permit limited, rather than full-scale, competition from new entrants for certain kinds of
services for which the common carriers had enjoyed
a protected market under the traditional regulatory
climate since 1934. Common carriers have met this
competitive element with a variety of tactics.
(See Appendix for the text of the "Bell Bill" proposal, designed to protect the natural monopoly of
common carriers.)
    The competition which has appeared recently is
of two kinds.
    In the first, which is traditional, new entrants
offer service where there is an unanswered demand.
These new entrants are much smaller than the common carrier telephone companies, dominated by Bell,
the world's largest telephone company, which has
87 percent of the telephone market and nearly all
the long-lines traffic in the country. Usually,
the new entrants propose to provide a new, but
limited, service which the telephone carriers are

either not providing or are providing at a rate
much higher than the proponents suggest.

The second kind of competition arises from the
use of alternative technologies by competitors.
The rapid development of alternative technologies
which can provide a particular service, let competitors take advantage of cost savings, or improved transmission techniques. This is clearly
an issue with undersea cables and satellites, and
land microwave systems and satellites. These cases
will be discussed later in this chapter. Technological changes, rapid recently in the telecommunications industry, present particular opportunities
and problems for this competitive setting. The FCC
is investigating the implications of competition in
the telecommunications industry with a fact-finding
proceeding.(6)

The issues that arise in this area of regulation
relate largely to the rates which the commission
permits the common carriers to charge for various
services. This complex issue has a long history,
but it can be summarized by saying that the commission has permitted, even required, the carriers to
use an average, per-mile rate for many of its services. Thus, without regard to the particular costs
the carrier encounters in providing a particular
service, the rate for that service will be uniform
for equal distances. This permits cross-subsidiza-
tion of services. In the heavily used portions of
the system, the rate is much higher than the cost
of the service. However, these high rates subsi-
dize the high-cost, low-use portions of the system.
New entrants providing various services can under-
price the common carriers by charging the cost of
the service, rather than an average for a nation-
wide system. This raises economic and political
issues for the common carriers and new entrants
about the costs of service and the rate of return.
In addition to the issue of rates, there can be
issues about the services provided and those de-

manded of the carriers. New entrants may try to
provide a service which the common carrier is not
supplying and for which the new entrant finds a
demand. The debate over services is illustrated
by several of the cases discussed in this chapter.

Technology and Regulation

There is an important set of relationships between
technology and regulation. Regulation frequently
involves control of technological development.
Various theories, some with empirical support, suggest the fashion in which a regulated industry will
develop technology and innovate new technologies.(7,
8) These theories generally point to the prospect
that an established carrier will use the already
existing technology, in which it has already made
major capital investment, until the investment in
the technology is fully amortized and depreciated.
The problem with this strategy, from the perspective
of technological innovation is that the amortization
of these technologies is spread over long periods
ranging from 20 to 40 years. In this industry,
where a new technological development can appear
at any time, and they appear more rapidly than
once in 30 years, the old technology will not be
replaced nearly as fast as is possible. Telecommunications is capital intensive, which means that
the physical plant requires great capital before
any service can be provided or any revenue is received for the service. The regulatory policy determines the company's revenue on the rate base,
which is a fixed percentage return on the investment by the carrier. Thus, the carrier has an
economic incentive to invest as much money as possible in plant and other rate-base items, so that
the fixed rate of return will produce the greatest
amount of revenue. Coupled with the capital intensive nature of the industry this incentive may
retard the adoption of a new technology or a cheaper technology.

The development and adoption of a new technology would be based on market mechanisms if there was no regulation to contend with. Under this scheme, the cheapest, most efficient technology would be used for a service and would be replaced when a cheaper and more efficient technology was available. Given the protected position of the regulated industry and the capital incentives to choose the most costly technology, the innovation rate within the telecommunications industry may be slower than in a free-market setting.(8) The actual rate of innovation and adoption is difficult to determine, and the evidence of innovation in telecommunications is mixed on this important point. However, it is important to remember that the rate of adoption and the direction of technological development depend in part on the regulatory climate in which the industry must operate.

Communication Satellites

The first serious suggestion that satellites in geosynchronous orbits (i.e., stationary above a specific point on the earth's surface) be used for two-way telecommunications was made by Arthur Clarke, a British engineer, in 1945.(9) He envisioned a satellite orbiting the earth at exactly the correct altitude (42,000 km or 22,300 miles) so that its period would be precisely 24 hours and it would therefore appear stationary above the same spot on the earth's surface. Unlike other heavenly bodies it would neither rise nor set but would remain fixed in the sky. By providing receiving and transmitting equipment on board the satellite, it could act as a relay or repeater between points on the earth that were out of line-of-sight contact with each other but were each able to "see" the satellite. Clarke envisioned the satellite providing a vast array of services. Among them were telephone, radio, television, and postal services.

He even proposed that individuals be able to utilize the relaying capability of the satellite
directly through their own private antenna systems
and appropriate transmitting and receiving equipment.

Much of Arthur Clarke's vision has come true but
many years have elapsed since it was first proposed
as a technologically feasible advance in telecommunications technology. Why was there such a long
time interval between proposal and actual development of telecommunication satellites? By what
route was this development actually carried out
and what societal forces have affected this development? What impact have communication satellites
had on society? These questions are the subject of
this section.

Satellite Technology Development. (10, 11, 12)
The launching of the Russian Sputnik I on October
4, 1957, ushered in the era of the satellite. Until then the pace of satellite development in the
United States had been slow. As part of a world-
wide scientific program, the United States had
agreed to launch an artificial satellite; however,
the launch vehicle development was being carried
out on a low-budget program independent of the
military rocket development program. The orbiting
of Sputnik, eerily beeping from outer space, provided a new impetus that catapulted the United
States into a space race ultimately leading to
successful landings on the moon. Along the way,
however, a whole array of communication satellite
technology was developed that has had a marked
effect on the kinds and costs of two-way telecommunications services that are available now and
that are likely to be available for decades.

The U.S. first conducted communication experiments with artificial earth satellites in 1960 as
part of the ECHO project. This was a joint program
of the National Aeronautics and Space Administration (NASA), the Jet Propulsion Laboratory of the

California Institute of Technology, and the Bell
Laboratories. It utilized a large metalized bal-
loon that served as a passive reflector. Radio
waves were bounced off the satellite's surface
allowing transmission between widely separated
points. Although the use of passive satellites has
many attractive features, its major drawback was
the elaborate ground station equipment required to
get a sufficient signal-to-noise ratio to permit
efficient communications. The received signal
strength is reduced proportionately to the fourth
power of the distance for passive satellite communication as compared with the square of the distance for active satellites. As soon as suitable
space-qualified power sources and attitude stabili-
zation systems became available, all experiments
with passive reflecting satellites were dropped.
    There were other drawbacks. The rockets used
to launch early satellites could place only relatively modest payloads into earth orbit. Typical
orbits were on the order of 150 km (93 miles) above
the earth's surface and had orbital periods of
about 90 minutes. Such satellites were not completely out of the earth's atmosphere and therefore
were subject to drag that shortened their useful
lives. Because of the rapid rate at which such
satellites move overhead, it is necessary to track
them electronically and to have a new satellite
come over the horizon before the previous one has
disappeared from sight. Experiments by AT&T with
the TELSTAR (1962-63) and RELAY (1962-64) satellites
proved the feasibility of active satellites at low
and medium altitudes, and communication systems
based on as many as 50 orbiting spacecraft were proposed for operational purposes.
    For a variety of technical and political reasons,
the low altitude satellite system that AT&T proposed was not adopted in the early 1960's. In fact,
an act of Congress excluded AT&T from directly participating in the development of satellites for

transoceanic communication, giving sole responsibility for such developments to a new company, the
Communications Satellite Corporation (COMSAT).(13)
Under the direction of COMSAT, the emphasis moved
from low altitude satellites to the geosynchronous
orbit proposed by Clarke in 1945. Achieving and
maintaining satellites in such orbits proved to be
a substantial technical challenge. However, successful launches as part of the SYNCOM series in
July 1963 and August 1964 established their feasibility. Great sophistication of the control and
station-keeping operations both on the ground and
in the spacecraft are required to achieve and maintain a geosynchronous orbit in the presence of the
many disturbances ranging from solar winds to grav-
itational perturbations due to the moon and the
planets. These problems have been solved and today
the geosynchronous orbit is a state-of-the-art
capability.

In 1964 the International Satellite Communications Consortium was formed by 14 nations interested in seeing that the development of international communication satellites was carried out
expeditiously. The United States' representative
to the international group was COMSAT. The first
commercial operational communications satellite was
launched from Cape Kennedy on April 6, 1965, and
was placed in geosynchronous orbit over the Atlantic Ocean to provide communications between Europe
and North America. This was INTELSAT I which provided 240 two-way-voice circuits. This satellite
was followed by similar satellites: INTELSAT II in
1967 also having 240-voice circuits but with increased signal power; INTELSAT III in 1968 having
1,200-voice circuits; and by INTELSAT IV in 1971
having 5,000-voice circuits. The satellite con-
figuration proposed for 1980 has a 25,000- to
50,000-voice-circuit capacity.(10) Four operational
and four spare satellites of the INTELSAT IV series
are in orbit. Two are over the Atlantic Ocean, one

over the Pacific Ocean, and one over the Indian
Ocean. The earth portion of the INTELSAT system
consists of 150 stations in 80 countries. The
global nature of the INTELSAT system is shown in
Figure 5-4.

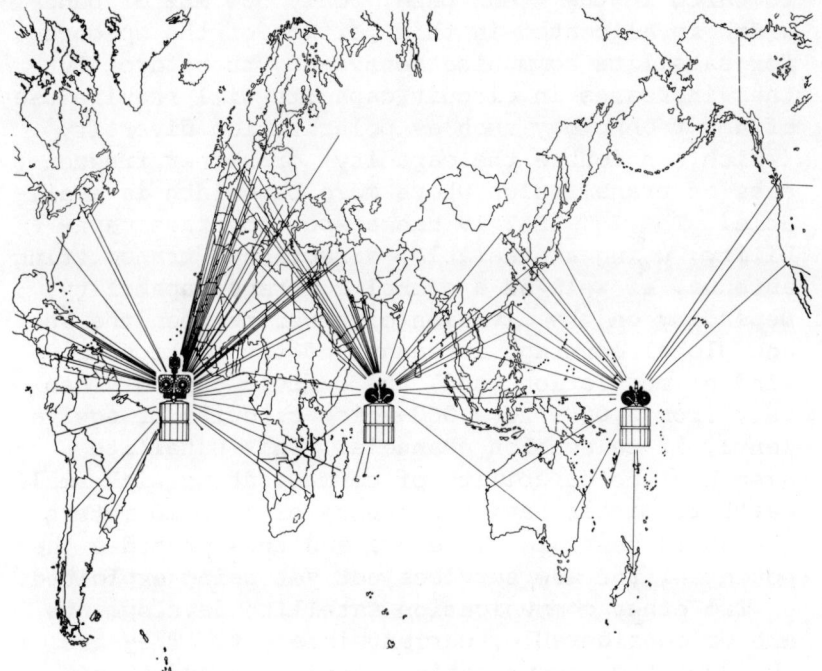

Figure 5-4 INTELSAT global system. *Science
            Magazine* photo.

The early INTELSAT series satellites were limited in channel capacity primarily by power, i.e.,
they were not able to transmit simultaneously on
all of the channels that were theoretically possible with their system bandwidth. For example,
the INTELSAT III satellite with a system bandwidth
of 450 MHz had a 1,200-voice-circuit capacity,
whereas the INTELSAT IV satellite, with a system
bandwidth of 500 MHz, was able to achieve a 5,000-

voice-circuit capacity because it had six times the
power of INTELSAT III. The INTELSAT IV is the
first of the systems that is bandwidth limited
rather than power limited.

Transmission to the satellites from earth is in
the 6 GHz frequency band and from the satellites
to earth in the 4 GHz band. Only 500 MHz of bandwidth is allocated in this portion of the spectrum
for satellite communications, and, therefore, further increases in circuit capacity will require use
of new technology such as polarization diversity
(which can double the capacity) or higher frequencies of transmission where more bandwidth is available. The INTELSAT IV has a spot coverage capability, using a steerable, directional transmitting
antenna, as well as a global coverage capability.
Depending on the division of power between the spot
and global coverage antennas and the particular
kind of modulation used, the circuit capacity can
vary from 3,000- to 9,000-voice circuits or equiva-
lently 12 television channels. This satellite
also has the capability of communicating with small
earth terminals having antennas of 5 to 10 meters
(16 to 33 feet) in diameter, and thus provides the
potential for new services not yet being exploited.

Two other communication satellite developments
are of considerable, current interest. They are
the domestic communication satellites and an experimental services satellite, the ATS-6, that is
demonstrating new potentials for satellite communication.

The first domestic satellite system to be placed
into operation was TELSAT-Canada's ANIK 1 which was
launched on November 9, 1972, and followed by ANIK
2 launched on April 20, 1973. In 1974 Western
Union placed two WESTAR satellites in orbit. These
four satellites are all similar in design and each
provides approximately 7,000-voice channels or 12
color television channels. Two new systems, each
utilizing three satellites, are being put into

operation by RCA Globecom and by AT&T and GTE Satellite Corporation. The initial satellites were launched in 1976 and employ polarization diversity to double their circuit capacity to 14,000-voice channels each. European domestic satellite development is proceeding but lags well behind that over the North American continent. In the late 1960's Russia deployed a satellite communication system based on low altitude (i.e., nonsynchronous orbit) satellites known as the MOLNIYA series. Not until 1974 did the USSR launch its first synchronous orbit communications satellite.

The rate of growth of available satellite voice circuit capacity has been spectacular, going from 240 in 1965 to over 50,000 in 1975. At the same time the investment cost per circuit per year for the INTELSAT IV is $1,000 and that for the next generation of satellite (1980) will be $100.(10) The cost and availability of long distance transmission facilities is no longer the primary constraint on expanded two-way telecommunications.

Applications Technology Satellite. On May 30, 1974 the Applications Technology Satellite (ATS-6) was launched from Kennedy Space Center and was placed in a geosynchronous orbit. This satellite differed from earlier spacecraft philosophy which had been directed toward minimum weight and maximum reliability in the spacecraft system.(14) In the ATS-6 the design philosophy was to incorporate sufficient sophistication into the spacecraft system so that low-cost, simplified ground terminals could be used to communicate to and from the satellite. The key feature of the ATS-6 is a 9.1-meter (30-foot) steerable antenna which provides adequate gain so that ground terminals with a 3-meter (10-foot) antenna can receive direct broadcasts of high quality color television signals. Also, the satellite was designed so that it could be moved to more than one geosynchronous orbit.

Initially, the ATS-6 was placed in orbit to serve North America including Alaska. A number of

experiments were carried out during the first year
to investigate the feasibility of relaying programs
through the satellite directly to such facilities
as schools, CATV systems, and clinics.(15) The overall telecommunications system to support these experiments included 120 video receive-only terminals,
51 radiotelephone transceivers, and eight video
originating terminals. Very promising results were
achieved in all experimental programs ranging from
transmission of education programs for rural schools
in the United States Appalachian region to small
medical clinics transmitting diagnostic video in
the Alaskan bush to a hospital in Fairbanks. In
May 1975 the satellite was moved to a new orbit
where it began providing direct television broadcasts to rural villages in India as part of a
further experiment in providing educational opportunities to regions having inadequate telecommunication facilities.

The experiments carried out with the ATS-6, the
world's most powerful and versatile communications
satellite may well herald a new era of technological aid to education, medical, and telecommunications services.

Political Arena of Communication Satellite Development

In the pre-1962 years, some government officials
and private citizens were concerned about how the
United States would develop and use satellite technology. At that time only the Soviet Union and the
United States were capable of launching artificial
satellites. Some private companies such as AT&T
were active in research and system planning of satellite communication systems and were considerably
set back by enactment of the Communication Satellite Act of 1962.(13) In his recommendations to Congress regarding satellite communications, President
Kennedy specified that to promote international
cooperation and to insure open and equitable access
to such communications facilities, a single corpo-

ration should be created. The law produced the
Communications Satellite Corporation (COMSAT), and
gave it a unique position in communications.(16)
   COMSAT was required to offer half its stock to
the public and the remainder to U.S. companies engaged in international communications. This meant
that the companies, which stood to lose the most
from the creation of a new entrant, received ownership as an economic incentive for supporting its
development. COMSAT was to have exclusive control
over the launching and use of U.S. communications
satellites that would be used for international communications. Interestingly, the initial view of
satellite technology was that its only practical
uses would be for *international* communication.
Clearly, from the 1962 act, everyone understood
that COMSAT's control would only reach to the international dimension of such communication. Besides permitting the international carriers to
purchase 50 percent of the stock in the new company,
they were also allowed to appoint six of the 15
members of COMSAT's board of directors while the
public owners appointed six and the President appointed three. This meant that the communications
companies (largely AT&T) would have a major voice
in the policy decisions of the corporation. This
created some conflict-of-interest questions which
were solved only by FCC ruling that COMSAT could
only serve the carriers and not compete with them
for service to users.(17) COMSAT thus became a
carrier's carrier.
   The technology implications of the 1962 statute
are highly significant. The new corporation was
not expected to develop the technology required.
However, the legislative history indicates that
COMSAT was to encourage technological development
through its acquisition of equipment for its systems and thereby provide hardware developers with
incentives to remain in the field. Major developers and manufacturers would have no significant

role in the development of satellite technology
unless they were able to obtain a contract from
COMSAT to produce such equipment or systems. Furthermore, the selection of the particular technology to be utilized was COMSAT's responsibility.
An example of this was the choice of the geosta-
tionary orbit system over the random orbit, low
altitude system AT&T strongly supported.

The act also created an interesting set of problems because the international carriers such as
AT&T and ITT could own or participate in the ownership of undersea cables while they were prohibited
from owning the satellites. Interestingly, many
major advances in undersea cable technology came in
the mid 1950's just before satellites became a
reality, and the international carriers began making major capital investments in such cables from
the outset. Technologically, satellites will unquestionably surpass cables in circuit capacity,
switching capabilities, and, with the exception of
transmission delay, quality of service. A study
by the National Academy of Engineering has estimated that for systems of up to 720-voice circuits
satellites and cables are economically competitive
but that for large systems the satellite system is
likely to be one or more orders of magnitude less
expensive on an investment-per-circuit-per-year
basis.(18) However, at present they are in close
equivalence for many types of service, and a potentially competitive situation between the two has
developed.

Competition is completely controlled by the FCC
which licenses transoceanic cables, as well as satellites. The effects of this regulation may not be
economically optimal, since the FCC has tried to
satisfy both sides of this effort by licensing both
facilities, sometimes permitting substantial duplication of services. For example, the FCC licensed
a new 720-voice-circuit cable from Florida to the
Virgin Islands in 1966. During the same year, the

FCC approved the building of a communication satellite earth station in Puerto Rico. Both the satellites and cable interests supported *both* services
even though one or the other of the systems would
appear to be redundant. The FCC stipulated that
since the then current demand for service would not
support both systems, the new traffic was to be allocated 50/50 between the satellite and cable circuits.(19) The reasons for this allocation appear
to be more political than economic or technological.

While domestic satellites were initially seen as
unlikely, subsequent technical developments have
caused the emergence of domestic communications
satellites as a major, current regulatory issue.
The FCC began inquiring into the development of domestic satellites in 1966, and only recently has it
reached a final policy outline for allocations.(20)
The current technology is well advanced and provides for attractive switching and circuit features.
Furthermore, it is clear that domestic satellites
will compete economically with a number of services
currently being supplied by other means.

Technopolitics of Communication Satellites

The politics of domestic communication satellite
development involved interests which sought either
an open-entry policy by the government, which pos-
its marketplace competition, or a regulated industry policy, which would restrict entry. The play
of political forces in this contest illustrates
much about the technopolitics concept. In the
early 1960's, the President (Kennedy) strongly supported a fully controlled satellite program. In
the mid 1960's, the President (Johnson) and his
task force did not support any particular policy
because of many unanswered technological and economic questions.(21) However, in 1972, the President (Nixon) supported an open-entry policy, and
the chairman of the FCC (Burch) also favored such
an open-entry policy.(20) COMSAT and the FCC staff

supported a controlled-entry policy, and there was
some Congressional support for this. The final result is that the FCC is operating currently with an
open-entry policy, licensing competing systems as
they are proposed. This may produce significant
changes in the development of communication satellite technology. New companies, new system requirements, and new technology will enter the field.
TELSAT Canada, Western Union, and AT&T-GTE have
already placed operational satellites in orbit; and
a consortium headed by IBM is seeking entry into
the field in the near future. It appears probable
that significant advances in technology and services will occur as a result of this competitive
setting for domestic satellite system development.

The past 12 years of political developments in
the satellite communications area illustrate different effects of political control on development
of telecommunications technology. The conclusions
are intriguing although not altogether definitive.
The political alliances change over time. The
alliances also change as the economic scene changes.
Thus, the private carriers took a much different
tack after the enactment of the statute in 1962
with regard to communication satellites. However,
the current domestic satellite picture seems to be
economically quite encouraging for them, and they
are active in a variety of governmental arenas
seeking to utilize their political advantages to
insure their technical position in this newly
emerging segment of the communications industry.

The alliances also depend on who the particular
office holders are, as in the case of three presi-
dents (Kennedy, Johnson, and Nixon) pursuing widely
differing policies on related matters in the course
of 10 years. The Federal Communications Commission
appears to be caught in the middle of much of the
decision making process--forced to contend with
diverse interests and conditions and apparently
trying to give something to each interest in each
set of circumstances.

The technopolitical model of the COMSAT decision is shown in Figure 5-5. The principal participants are the new entrants (international
carriers), the Executive, and Congress. The new
entrants were resisting the closed-entry policy
that ultimately prevailed while the Executive supported this policy. The final allocation was made
by Congress when it passed the Satellite Communications Act of 1962. The principal result was to
concentrate responsibility for technological development in a single organization, COMSAT.

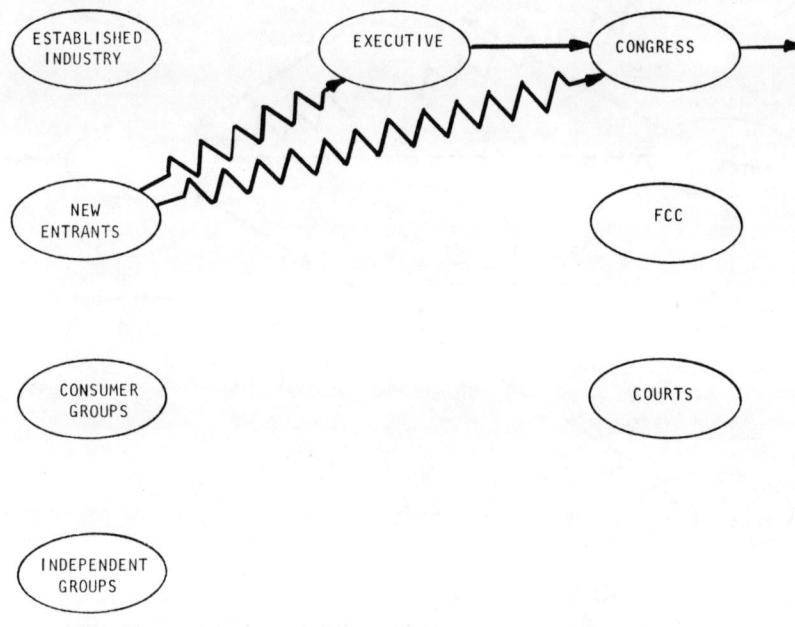

Figure 5-5 COMSAT allocation model
New entrants include the established common carriers
    seeking entry into the ownership and operation of
    communications satellite facilities.

The model for the domestic satellite decision
(DOMSAT), Figure 5-6, is substantially different
than that for COMSAT. Here the established in-

dustry (COMSAT) opposed what turned out to be the
eventual policy while the new entrants (common and
special carriers) and an independent group (Ford
Foundation) supported it. Congress took no firm
position, and, therefore, the policy of open-entry
proposed by the Executive prevailed. The result of
this allocation process was to introduce significant competition into the technology and service
aspects of domestic satellite system development.

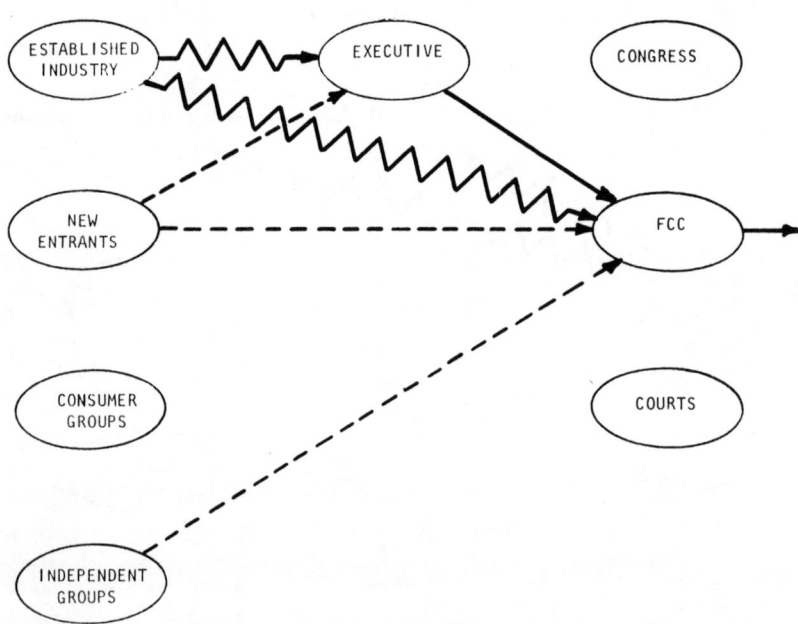

Figure 5-6   DOMSAT allocation model

Established industries include the established common
    carriers, particularly COMSAT, who sought to re-
    strict entry into domestic communications satellite
    to those experienced carriers.
New entrants include Western Union, ITT, IBM, and RCA
    who sought entry into the domestic satellite busi-
    ness.
Independent groups include the Ford Foundation which
    sought to foster open competition.
The Executive also sought to foster an open competition
    policy for domestic communications satellites.

Foreign Attachments and Carterfone

The Bell Telephone system has long had a policy of
discouraging interconnections with nontelephone
systems and with non-Bell telephone devices. The
refusal of Bell to interconnect with other telephone companies in the early decades of this century led to anti-trust charges and litigation which
ended in the Kingsbury Commitment of 1913 after
which Bell willingly interconnected with competing
telephone systems to avert possible antitrust
action by the government.(5) In addition, Bell has
long been able to prevent the attachment of foreign
devices to its telephone system, by requiring that
anyone who has a Bell telephone agree not to make
such attachments.(22) Bell reserved the right to
disconnect or remove any such devices if they were
attached by a telephone user.

This policy did not prevent manufacturers of
equipment from developing and selling various devices to be attached to phones in order that the
user could enjoy a service or advantage which the
telephone system itself did not provide. The FCC,
through the *Hush-A-Phone* decision in 1955, upheld
Bell's prohibition of foreign attachments.(22)
Bell's argument for such authority rested largely
on the need to maintain the integrity of the telephone system, which Bell claimed it could not do if
various, non-Bell devices were tied into the system. Bell argued that they were responsible for
the quality of the service to their customers, and
they could not guarantee that quality if such devices were permitted on their phone lines. The
Federal Court of Appeals reversed the FCC on this
decision, and that resulted in the first crack in
the otherwise complete prohibition against such
foreign attachments.(24) The Bell companies inter-
preted the decision as narrowly as possible, and
generally refused to interconnect after that.(5)

The second major effort to achieve interconnection for foreign attachments was successful,

when in 1968 the FCC held in *Carterfone*, that the
Bell system had to allow interconnection, as long
as the integrity and quality of the operating system was maintained.(24) (See Appendix). Thus the
Bell system filed new tariffs with the FCC listing
the charges for such attachments, and the requirement that the attachment be made to the Bell system
with a device produced by Bell. Although they
argued that technically it was costly and difficult
to make such attachments, Bell was able to provide
the interconnection device quickly after the FCC
authorized Carterfone attachments.

Carterfone greatly opened the market for foreign
attachments. These devices included such items as
recorder-answering systems, burglar alarms, local
switching and intercom systems, and a host of additional, specialized products. The manufacturers
and developers of these devices certainly benefited
from the *Carterfone* decision. Bell was required to
make a major adjustment in its attachment policy
and in its technological development. Currently,
the device necessary for attachment is being produced under Bell license by non-Bell companies and
is installed in the actual foreign attachment
rather than on the telephone line after the device
has been purchased. Thus, the cost of such interconnection is included in the price of the attached
device, the Bell's fee is a license fee paid by
the equipment manufacturer.

Although the AT&T system could certainly capi-
talize on the vast and increasing market for such
services, it chose not to provide such systems or
services, except local exchanges. It has settled
for an attachment fee for the use of the Bell-
designed attachment device, which must be incorporated into all foreign attachment devices. The
telephone company appears to have ignored a major
service market that is currently developing, and
which the company did not service for a number of
years while it pursued its absolute ban on foreign

attachments. It seems that either Bell could not
compete in price with other manufacturers in this
field, and therefore did not respond; it chose not
to try because the *Carterfone* decision caught Bell
off-guard and unprepared to compete in that market;
or it chose to make its capital investments in
other areas. Most observers would define this market as a portion of the natural monopoly area of
the common carrier, but it has not been absorbed
by the monopolist in this case. As a result, a
large number of small manufacturers are competing
in this field, and the optimality of this arrangement is unclear. However, the development of new
devices and technologies is proceeding quite
rapdily.(22)

The model of the *Carterfone* allocation decision
is shown in Figure 5-7. Here the demand of the new
entrant (*Carterfone*) was supported by the FCC over
the strong opposition of the established industry
(AT&T). Bell voluntarily dropped its initial appeal to the courts, and it filed new tariffs with
the FCC within six months of the *Carterfone* decision. The net result of this decision was the
strong encouragement of technological development
in an area previously having very limited market
potential.

Specialized Carriers and Competition

A characteristic of the telephone company is that
it is a common carrier, which means that it provides service to all comers, upon demand, at a
regulated rate. In the past 20 years a group of
carriers has developed using the same or different
technologies to provide selected or specialized
services to particular users. These services derive largely from the microwave link systems which
developed after the microwave technology was devised during World War II.(5) The Bell system has
a monopoly on long-distance transmission of voice

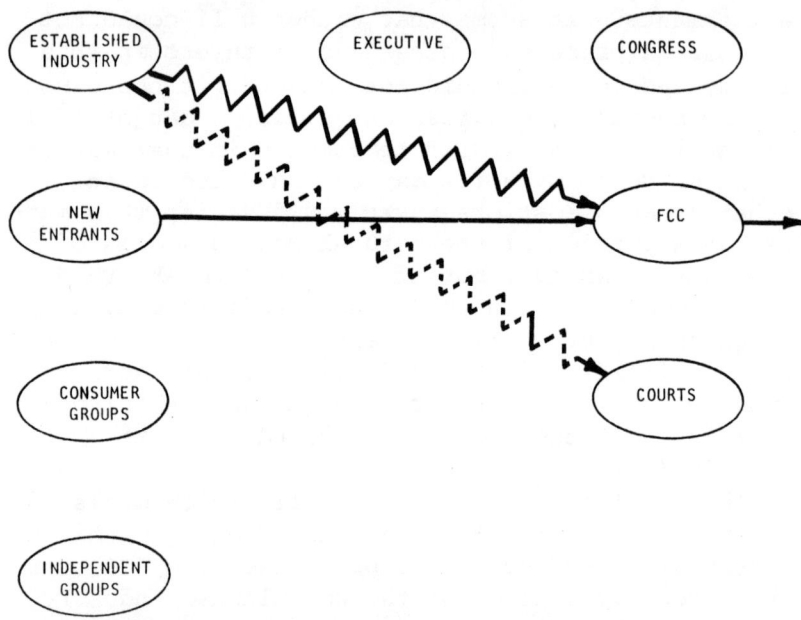

Figure 5-7   Foreign attachment model

Established industries includes the Bell Telephone system which sought to convince the FCC that foreign
   attachments were not necessary and were injurious
   to the telephone system.   This same argument was
   presented to the courts for a short period of time.
New entrants included Carterfone, as later many other
   equipment manufacturers, who argued that such
   attachments were permissable and noninjurious to
   the operation of the telephone system.

communication over microwave links in this country.
It has developed an extensive microwave system, and
nearly all the long-distance telephone calls made
in the country (37 million per day) are carried
over this system.(26)

   New entrants, willing and able to provide similar microwave based services, began appearing in
the late 1950's. The new services and the lower
rates which they proposed would require more of
the electromagnetic spectrum than was then allo-

cated for microwave systems. The initial question
was whether adequate spectrum was available to
accommodate these competitive requests, or whether
the economies-of-scale argument would support the
monopolist's claims of the AT&T system that only
it should be allowed to use the microwave frequencies for long-distance communications.(27)

The Above 890 Decision

In 1959, the FCC made a major break with established policy by its *Above 890* decision, in which
it held that the spectrum above 890 MHz was large
enough and technically open for use by other carriers using microwave links.(28) This was a first
step through the door for other potential carriers,
but little came of it at the outset since there
were no applicants to use the spectrum above 890.
Most of those advocating that the commission open
up that portion of the spectrum for microwave use
were not interested in becoming carriers but
rather would use such proposed services or would
benefit because they manufactured equipment necessary for non-Bell carriers to utilize that portion
of the spectrum.(29) This is an interesting point,
since the potential users were in the background
during the early fighting to open up the spectrum
and appeared only some years later with specific
proposals for using the spectrum. The battle was
initiated by equipment manufacturers and customers.
    The Bell system responded to the *Above 890* de-
cision by providing services which it had not provided earlier.(30) These services included Telpak
rates for the rental of data transmission lines in
smaller sizes than the Bell system had provided
earlier, and at rates low compared with Western
Union's Telex system. In fact, the FCC eventually
disallowed some of the lower Telpak rates as predatory after formal complaints had been made by
Western Union.(31) The Wide Area Telephone System
(WATS) lines were devised in response to the threat

of competition during this period, again in an effort to provide a marketable package, at an attractive price to customers. This suggests that once
the common carrier monopoly was faced with potential competition, it began changing the product it
provided.

The MCI Decision

The first decision of the FCC which licensed a new
specialized carrier in the above 890 MHz portion
of the spectrum was the *MCI* decision in 1968.(32)
(See Appendix). MCI proposed to provide only a
microwave link between St. Louis and Chicago,
nothing more. The user would have had to supply
its own distribution link at either end of the line.
The major attraction of the MCI proposal was that
the cost of the service was much lower than that of
AT&T. Furthermore, the MCI service was designed
for small users who needed the communications link
only occasionally or for short periods of time.
Thus, the market MCI sought to reach was comprised
of low volume users which the Bell system did not
adequately serve. The FCC approved the MCI application for construction permits over the loud ob-
jections of AT&T which argued that MCI's proposal
did not provide a new service, that it would be
a nonoptimal use of the radio spectrum, and that
MCI was not financially responsible enough to en-
trust with the building of the link. The commission based its decision on the prospective customers which MCI had exposed by market research, the
economic position of the company, and the fact that
the proposed service was indeed different from that
provided at AT&T because it reached a market which
the common carrier had not recognized earlier and
which it certainly was not serving when MCI made
its proposal.

Although the FCC was careful to state in the
MCI decision that that decision would not be considered a binding precedent for future applica-

tions of specialized carriers, the *MCI* decision
prompted a flood of proposals from other entrants
in the area of specialized carrier services. In
fact, during the year after the *MCI* decision, proponents made 37 applications, involving more than
1,700 microwave stations--over one-third as many
stations as the entire Bell system had.(5) The Data
Transmission Company (DATRAN) made the primary
proposal for a nation-wide, switched network to be
used solely for data transmission. The DATRAN
proposal offered complete data transmission service, which would expend some $350 million.(33)
    The DATRAN proposal was significant because it
offered to provide a data transmission service
which the Bell system did not then provide, and it
offered to establish, at great expense, a nation-
wide microwave system to do it. The data transmission business arose during the 1960's as more computers developed and the networking of computers
became feasible and highly beneficial to users.
The Bell system did not provide any major data
transmission services, other than to permit the
leasing of a voice-grade telephone channel over
which the user could choose to transmit data. This
was expensive and required the user to tie up much
more channel capacity than necessary to transmit
data.
    Bell's response to the proposals for data trans-
mission systems was a major effort to develop the
technology to permit it to transmit data quickly
and cheaply.(29) It came up with a system it called
Data-Under-Voice. This required minor adjustments
at low cost to the existing microwave system Bell
used for long-lines communication. Bell's data
competitors sought to prevent the use of Data-
Under-Voice at the price Bell proposed because it
was much cheaper than that which they had to charge
to recover the costs of building an entire system.
(34) But Bell did not develop the Data-Under-Voice
system technology until challenged by new entrants,

who sought to provide this service. Bell either
felt little need to service the data transmission
market, until competitors tried to enter, or it
simply did not choose to provide a means cheaper
than voice channel in order to service the demand.
The technology Bell developed in response to the
threat of data transmission competition involved
a major effort at Bell Laboratories--almost a crash
program. It began the program in earnest after new
entrants presented a competitive threat to Bell's
monopoly position.

Bell also responded to the challenge of specialized carrier competition by lowering specific rates
on service for which competition appeared. As
noted previously, the Bell system uses rate aver-
ages so that some services return revenues much
higher than the costs of that service while others
do not "pay" for themselves. If Bell were able to
identify the costs of a particular service and then
charge a rate according to those direct costs, some
of their services would cost a user much less than
the averaged rate, and Bell sought to charge this
actual-cost rate for some of these services. The
new entrant-competitors challenged these pricing
practices as predatory and as contrary to the rate-
averaging, rate-of-return structure which Bell has
followed as a regulated industry for some years.
The Federal Communications Commission invalidated
some of Bell's rates on the grounds that they were
not average, and the direct costs of the service
could not be identified clearly enough to warrant
such pricing of the services. Bell argues that if
they are to be faced with competitive new entrants
for some services, which involve "cream-skimming,"
according to Bell, then the common carrier should
be permitted to *compete* with the new entrants.(34)

Problems in Competition

The injection of some competitive elements into the
services of the common carriers raises important

economic and political questions for the regulator.
First, if it is beginning to consider competition
as a tool for regulating an industry, should the
entire regulatory scheme be examined with an eye
to making all necessary adjustments so that competition achieves the desired results? The regulator
will have to articulate what its objectives are and
what policies it is seeking to implement by permitting new entry. Then it will have to spell out how
these objectives will be achieved by the competitive efforts.

The second question about this competitive orientation is how can the regulator identify the extent of the natural monopoly which a common carrier
should enjoy?(5) That requires outlining with what
services and in what forms the new entrants can
make a profit or contribute a service. The precise
boundaries of the services for which the economies
of scale operate are difficult, if not impossible,
to identify. Yet, it is in these areas that the
natural monopoly should exist. Recent efforts suggest that there is little likelihood of the new
entrants achieving economies of scale on intercity
telecommunications services.(35) Since its original
efforts in the specialized common carrier market,
DATRAN has failed financially. This is one of the
major, long-term problems with selective competition and economics of scale. A further part of
this question is, should competitors be permitted
to try to compete where the investment pattern is
highly capital intensive?

A third important question has recently arisen.
The new entrants are seeking protection from the
common carriers for their various rate structures
and services.(36) DATRAN asked the FCC to bar Bell
from instituting part of Bell's 24-city data transmission service which would directly compete with
DATRAN's. If the common carrier can provide identical or better service for less or equivalent
rates, should users be prevented from choosing the

cheaper service--in a competitive market which the
FCC apparently seeks to create--because the new
entrants will not be able to survive economically
if competition really breaks out with the common
carriers? This policy question has various answers depending on what one's goals and values are.
However, if competition is to be used for regulatory purposes, the regulator must spell out and
follow a clear set of objectives unless the morass
of regulation is to increase and impede future
developments.

The MCI Decision Model

Figure 5-8 presents a model of the *MCI* decision.
The participants include the established industry
(AT&T), the new entrants (MCI), and the users or
potential market. The efforts of these people include active opposition by AT&T. The potential
MCI customers provided support for the MCI application which the commission partially relied on for
its decision. Of course the proponents (MCI) actively made and defended the request, which was
successful. The established industry sought court
reversal of this decision, but to date this effort
has been unsuccessful.

The Video Telephone

The development of the video telephone illustrates
several important considerations which have not
been discussed thus far. Primarily, the development of this device has not been directly regulated
by any government agency. In fact, this system has
largely resulted from private development and technical decisions, particularly by the Bell Telephone
system. This fact alone may explain some of the
unique technical developments and failures which
have occurred in the long history of this device.
    The Bell system began developing a video tele-
phone (the Picturephone®) in the 1940's and early

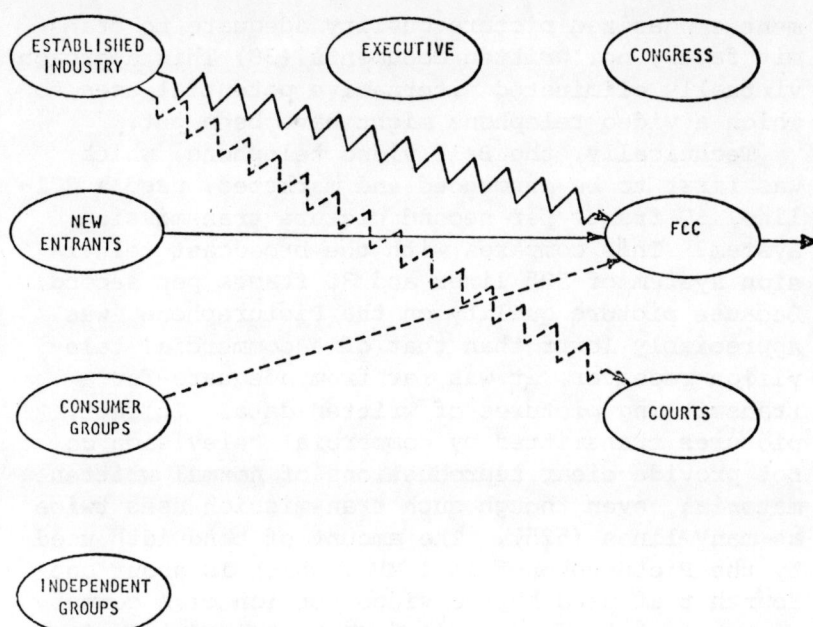

Figure 5-8 MCI decision model

Established industries included the common carriers,
   particularly AT&T, who argued that the proposed
   service of MCI would not meet a need that was not
   already being met by AT&T. The company later
   challenged the FCC decision in the courts, but
   lost.

New entrant was the MCI Company which sought to provide the competitive microwave link between Chicago
   and St. Louis to meet demands not then met by AT&T.

Consumers groups involved the potential MCI service
   customers which came forward to support MCI's application before the FCC.

1950's,(37) billing it as the logical extension of
the audio telephone. Bell's technical decisions
suggest that it intended to market the Picture-
phone® primarily as a face-to-face communication
device and did not intend to emphasize the other
potential uses of the video telephone. As a result of this marketing approach, the Bell develop-

ment emphasized picture quality adequate to transmit faces, not written documents.(38) This approach
virtually eliminated alternative potential uses to
which a video telephone might have been put.
Technically, the Bell video telephone, which
was first to be announced and marketed, used a 251-
line, 30 frames per second picture transmission
system. This compares with the broadcast television system of 525 lines and 30 frames per second.
Because picture quality on the Picturephone® was
appreciably lower than that of a commercial television receiver, it was far from adequate for
transmitting pictures of written data. In fact,
pictures transmitted by commercial television do
not provide clear reproductions of normal written
material, even though such transmission uses twice
as many lines (525). The amount of bandwidth used
by the Picturephone® is 1 MHz, which is about one
fourth that used by the video portion of a commercial television system (4.6 MHz). A bandwidth corresponding to nearly 100 voice channels is required
to transmit one Picturephone® conversation.(37)
This technical requirement would have an enormous
impact on the amount of telephone switching and
transmission equipment which the Bell Telephone
System would have to install in order to replace
the existing audio system with a video telephone
system. Since one video conversation would be
equivalent to 96 audio conversations, the growth in
plant would be astronomical in order to completely
replace audio telephones.
Bell's development of the Picturephone® continued until the late 1960's when it began first
marketing efforts in Chicago, Pittsburgh, and,
later, Washington, D.C. The service provided for
the video telephone was local only, so the subscriber could not use the device to make long
distance calls. Furthermore, in the initial stages
of marketing, few other people with Picturephones®
could be called. Most of the early users were com-

mercial establishments which used the Picturephone®
for interoffice or interplant communications, or
for customer shopping. The cost of the service was
much higher than for regular telephone service, because of equipment and installation costs, but apparently the cost to users was lowered in order to
make the device saleable. Even so, there seemed
to be little demand for the service, and the Bell
System has suspended marketing the Picturephone®.
(39,40) One former Bell Laboratory official has
indicated that the world was not ready for the
Picturephone®.(41) Other less favorable comments
suggest that the Picturephone® has been a "flop."(42)

It has been roughly estimated that Bell spent
hundreds of million of dollars in the research, development, and pilot production of the Picture-
phone®. Whether or not this is an accurate figure,
the amount spent has not returned any revenue.
While there may be explanations for such results,
there are no clear answers, and the ones provided
are tentative at best.

The first explanation is that Bell totally mis-
judged the market for such a device. One observer
has suggested that the estimated penetration of the
market could have been around five percent, certainly not a major portion of the telephone market.(37)

Second, it could well be that Bell failed to
estimate the psychological factors which might
inhibit people from using the device. However,
since the public never widely used the Picture-
phone® , psychological inhibitions, which do seem
to exist, could not have been the significant
factor.(43)

A third, and more probable explanation, is that
the system was designed to do little more than provide face-to-face communication. Perhaps, the
crucial need is more for a system which transmits
pictures of documents and other written materials.

The Ericsson Telephone Company of Sweden, an
international telephone equipment manufacturer, has

done much recently to develop a different video
telephone system. This company emphasized an entirely different orientation toward the functions
of the system and its viability. The Ericsson
Company has focused, not on face-to-face communication, but on transmission of written documents.(44,
45) It utilizes a system with a much sharper picture
than the Picturephone®. The Ericsson system involves 625 lines and 25 frames per second. This is
the European standard television system, which provides a sharper picture than American television.
The price of obtaining this sharpness is a substantial increase in the bandwidth required. The system requires 5 MHz for the transmission of its
picture and voice, nearly as great as the bandwidth
of a commercial U.S. television channel. However,
to reduce the cost of this transmission system,
Ericsson has so far used it only where the entire
system could be wired into a single plant. In pre-
liminary tests Ericsson installed the system in its
home office in Stockholm and used it among various
offices of the production, design, and order units.
Ericsson wanted to reduce the need for face-to-face
conversations where technical details had to be discussed and modified, and it found that the saving
in time and reduction in delays could pay for the
complete system in three years.(46)
    This use of the video telephone differs substan-
tially from the proposed uses of the Picturephone®.
It replaces internal conferences and upgrades interoffice intercom systems by transmitting draw-
ings and other written documents. While face-to-
face conversations are possible, the usage was
heavily oriented toward transmitting documents.
The picture resolution was quite high, compared
with the Picturephone®. The resolution of the
picture, an approach rejected by Bell developers,
makes the system valuable. Ericsson is beginning
to market the system to other users on a limited
basis, but, it is emphasizing use of the system
primarily within an office or organization.

The bandwidth requirements of the Ericsson system would place tremendous channel-capacity demands
on a telephone system for long distance service. To
make possible long-distance transmission Ericsson
has developed a slow-scan system.(45) It samples
the picture and transmits only those portions of
the picture which have changed (moved) since the
last sample. There is a time delay of some 30 to
60 seconds for transmission of a single picture.
However, this delay does not seem crucial in long-
distance transmission when the primary picture
usage is for documents. The slow-scan system allows the bandwidth to be compressed down to 1 MHz
which is the same as is used in the Bell Picture-
phone® or even down to 4 kHz which is the bandwidth for a regular audio telephone voice channel.
The long-distance use of this system has been
limited to experimentation.

The video telephone development experiences of
the Bell and Ericsson companies represent basic
marketing and technical differences, which may
lead to differences in the success of the two systems. Bell's philosophy appears to have been in-
correct, in that it did not achieve the desired
results. Ericsson seems to be quite encouraged
about the early success of its system, although
it is too early to confirm its "unqualified" success. The Ericsson system is based on a more
restricted and limited use of the video telephone,
and that approach may be the difference between success and failure. The video telephone appears to be
quite useful and successful for certain, limited
functions, but not as a replacement for the basic
telephone.

This technological endeavor and its results suggest that careful examination needs to be made before a technology is developed and chosen for a new
system. Many wonderful technological devices, possible within today's state-of-the-art, may never
be realized if careful market analysis and evalua-

tion of psychological parameters of the device are
not explored and properly taken into account.
These considerations are difficult, if not impossible, to carry out precisely. Yet, without
analysis and evaluation new technological innovations may be doomed. As the video telephone illustrates, such decisions can be made in various
ways, and the developer will have to live with the
consequences.

Summary

The development of integrated circuit technology
has made it economically possible to produce complex electronic components and systems. The modern
electronic computer, communication satellites, and
the electronic switching systems of the telephone
companies are typical examples of systems made
practical by integrated circuit technology.

The computer has also created a demand for a new
kind of telecommunications service--data transmission. The need for greatly expanded data transmission services has led new entrants to attempt to
enter the marketplace and to supply such services
between cities. This has raised difficult regulatory problems for the FCC since it must give the
basic authorization to provide such services.

The difficulty arises because of rate determination policies that have developed over the years.
Rates for each telephone company service are obtained by averaging over a wide variety of services
and locations so that they do not represent the
actual costs of that particular service. This has
made it possible to cross-subsidize certain high
cost services such as rural telephones and local
calls from revenues obtained elsewhere, e.g., long
distance calls.(47) If new entrants are going to
provide limited services at rates which reflect
only the costs of those services, their rates may
be very different from those that would have been

charged for the same service the telephone company
provided. In the case of data transmission, the
telephone company already has microwave links installed and could likely provide services at much
lower cost than a competitor, if only the marginal
costs of supplying the service were charged.

On the other hand, without a competitor the
phone company would probably structure its rates
differently and use part of the revenue from this
service to reduce the costs of other services. It
is this confusing area into which the FCC has
stepped, licensing of new entrants such as MCI and
DATRAN in the data transmission field, and authorizing competitive domestic satellite systems.

The pattern of regulation by the FCC has generally been that of supporting the established industry wherever its continuing business is concerned. This applies to broadcasting as well as
to two-way telecommunications. However, where new
services such as those stemming from use of foreign
attachments, intercity data transmission, and domestic communication satellites are concerned, the
FCC seems willing to allow a limited amount of
apparent competition. The degree to which a competitive situation actually develops will be determined by the methods of rate structuring the
FCC allows. Judging by previous performances all
of the companies will be required to charge the
same rate for the same service and there will be
little real competition. Therefore, the degree to
which society will benefit from the pseudo-competi-
tion authorized by the FCC is uncertain.

The switched telecommunications system in the
United States is dominated by the Bell System and
there is little reason to doubt that this situation
will continue to prevail for many years to come.
Bell has vast resources and is determined to improve and expand its services and to compete strongly with any new entrants in its marketplace. Because of the intricate nature of the rate-averaging

process Bell uses in determining costs of services,
extensive ramifications in costs of apparently un-
related services probably will occur when something
is done that forces a change in the cost of a particular service. There is considerable doubt
whether meaningful competition should or even could
be introduced into today's system. This is particularly important because causing a significant
degradation in certain services by such a process
might actually be easier than causing a significant
improvement. It must be remembered that the telephone service is generally excellent and nothing
should be done that might degrade it. A radical
alteration in services is not needed so much as
altering the directions of technological development to provide services and equipment more responsive to users' needs and wishes. This matter
is discussed further in the next chapter.

Chapter 6

Telecommunications' Impacts on Society

From earlier discussion in this book, the course of development of telecommunications technology and services varies with many factors. However, it is also important to consider what impact telecommunications development has on the society which uses the various devices and services this technology supplies. Precise discussion is not possible because the impacts are so numerous, so difficult to measure, and often evolve so slowly that they cannot be clearly discerned. One can hypothesize or guess about what effects technology has had or will have. Some individual technologies have provided clear and precise examples. However, to present a comprehensive picture of what effects telecommunications technology has had on American Society would be impossible. This chapter will examine several areas which have been documented and in which there have been or will be impacts. This will isolate selected topics and ignore other areas. Though the best treatment would be to explore the area systematically and completely, such an effort would require a large and extensive study, far beyond the scope of this book.

Some Areas of Impact

It would be useful to examine the impact of telecommunications technology by using some sort of

framework or approach so that the effects can be
placed in perspective. The framework which will
be used here is a general one. It consists of
grouping under appropriate headings major segments
of society which have been directly or indirectly
affected by telecommunications developments in recent years.

Politics and Public Institutions

The changes which telecommunications devices can
effect in this area are both promising and potentially dangerous.(1) First, it will soon be possible to take public opinion polls, almost instantaneously. These will give the elected public
officials some remarkably accurate readings of what
the public prefers on a policy or general subject
at any particular instant. This could be done by
attaching a two-way device to all sets on a cable
TV system. Readings could be tabulated in computers linked across the nation so that the officials could tell, by state, region, or individual,
what policy preferences were held on any issue.
This would profoundly affect the conduct of official business, since a basic element of a democracy
is that the wishes of the people be implemented.
Even competing interests make such a task difficult, and simply knowing precisely the wishes of
the people would greatly facilitate political decision making in which these interests are considered. This would make it difficult for a Con-
gressman or Senator to vote on a bill without at
least knowing what his constituents felt about the
issue, even if he or she chose not to support these
feelings.

This phenomenon, which we might call instantaneous democracy, would provide the decision makers
much information which they have to guess at or
have to pay large sums of money to discover now.
Such a system is not without its difficulties since

instantaneous voting or polling on issues might inject a lot of instability into the system. Thus,
if major policy issues were decided by voting or
polling in this way, much of the slack and delay in
the political system would disappear, and this
might cause the government to become less stable--
to the point that extremists or small groups could,
under emotional circumstances, control the government and drastically change what the political system did and how it did it.

To date, no one has advocated that such an instant polling system be introduced, either because
of apprehension over what effects this might have
on the operation of the political system or because
it is too costly. Instant referenda are not now
considered viable. However, the technology to
achieve them exists. Some public officials have
taken advantage of lesser technological advances
to tap, more closely than before, their constituent's views. The installation of a WATS telephone line between various offices in the district
and the legislator's office gives his supporters a
"free" chance to let the official know what they
feel on any subject they choose to discuss. This
means is not as representative or as instantaneous
as the poll, but it does give the representative
better information than was available in the past,
and at a fraction of the former cost.

In addition to such polls, the general telecommunications expansion permits political decision
makers, like others who must decide, to recall and
use vast amounts of data and data analysis which
used to be impossible. Thus, a legislator, or a
chief executive can use computers and large pools
of data before deciding many of today's policy
issues. This would not be possible without current telecommunications services and continually
decreasing costs. These facilities expand his perspective and his understanding of a policy problem
before he acts.

A major development in politics and telecommunications has been the development of news reporting,
particularly of elections by the television networks and other media.(2) A major television network can usually predict the outcome in statewide
elections on the basis of small numbers of votes
which are reported early in an election day. This
ability is due, first, to the development of sophisticated statistical techniques, and, second,
to telephone communications and computers. Whether
this projection of outcomes affects other voters,
who view them before voting, appears doubtful.(3)
However, processing this information and dissemi-
nating it has become nearly instantaneous. Possibly the television communication system is the
primary way people obtain their information about
political candidates, and the way they base voting
decisions, although this is not borne out by empirical studies. Again, empirical evidence suggests that the electronic media do not have much
impact with many people, though the potential is
there. This is due largely to the ability to communicate rapidly with many people.

A set of questions relating to telecommunications technology involves the government's use of
personal data and investigations of people.(4) The
data bank problem has been discussed widely in
recent years, because of proposals to create a
national data bank which would pool various existing sources of information on people. For example,
the FBI has vast files of known and suspected
criminals; the Social Security Administration has
a file on everyone with a social security number.
The collection of such information as living habits
and styles, health records, credit records, and
other kinds of data could, when pooled, produce
various kinds of "profiles" on people which might
be inaccurate, might be used for various unintended
purposes, and might seriously injure persons and
their reputations if not controlled and monitored
carefully.(5)

Whether the government will use such files, and
what sorts of controls the government will likely
place on the collection and use of such personal
data by agencies are important questions and problems which bear watching in the years to come. As
communications technology develops it will contribute greatly to the ability to create such files,
use such information, and control its usage.

Other considerations are important. One of the
most important involves the area of censorship and
the fact that the free press privileges in this
country are well established and have a long history. Where official, governmental agencies
establish or control information, the possibility
of a vast system exists to control more information. However, many independent broadcasters and
news sources reduce the likelihood of such censorship. If the government can control the channels
of communications and can determine what is presented, then the people hear only what the government wants them to hear. The situation portrayed
by George Orwell in *1984* may come true, since
modern telecommunications technology would permit
centralized and complete control of the channels
of communication.(6) There may be great protection
from this danger in radiated signals, since a government cannot as easily control what is broadcast
as it can what goes through a coaxial cable. If
every home has an optical fiber which brings all
communication services into the home, control of
that system by government would be easy. However,
if parties are permitted to broadcast over-the-air,
establishing such control is not as easy. Some
governments in the world control radiated signals
closely and seek to provide the user with only one
view of the world--that government's view. However, this is more difficult when there is no
single point in the communications system through
which all communication must flow. Such a situation will not likely develop in this country, but

the technology permitting such control of information exists. A safeguard against such developments
is the use of multiple information sources and
multiple technologies.
Consider that technology would permit new and
different forms of propaganda broadcasts. Thus,
closed-information countries now use various technical means to jam broadcasts from foreign governments or other propagandistic sources. However,
communications directly from satellites to home
receivers could happen.(7) If it did, the target
government would have to develop new ways of inter-
cepting and blocking such transmissions. The likelihood of this occurring and the need for counter-
measures depend on the political situations in
various countries. Obviously the occurrence of
this in some countries today would be strongly
opposed and in other countries might prove beneficial. The technology is certainly available,
but whether this develops, how it would be used,
and what the precise results would be depend on
the political situation in various countries.

Commercial Transactions and Institutions

The way in which business is conducted depends
largely on telecommunication services and, as time
goes on, that dependency will increase. Today's
commonplace ability to dial any telephone in the
country, directly, makes the conduct of some businesses much easier than in previous times. The
ability to have a continuous communications channel with a production plant thousands of miles
from the main office, permits a company to monitor
closely the conduct of its business, reducing the
costs of unforeseen delays and increasing profits
by having sufficient inventories available at appropriate times. Besides being linked by a telephone system, a company can use devices in its
internal communications system such as the video
telephone, services totally impracticable 10 or

even five years ago. In addition, more traditional intercom systems and local switching systems greatly facilitate communications with customers, supervisors, and anyone else who is concerned with the business.

The development of new devices to do old tasks,
such as computers to do accounting, can provide
incentives for companies to use such devices to
accomplish even more. Thus, the company may begin
performing new services or functions, even though
these were not anticipated or intended at the outset. The capability of a central office to have
instantaneous control of production plants and
warehouses all over the country means that businesses can be run more efficiently and with greater
ease than was possible a decade ago. Banks can
now conduct many of their customer-related transactions without need for a teller, since, electronically, all the necessary information can be
provided by the user at remote terminals. The
concept of a cashless society, founded on the
plastic credit card, is more a reality than a
speculation. The monitoring of credit cards--
checking validity--is quite simple, and the transaction in many stores occurs instantaneously, so
long as the customer has his card with him. What
these procedures permit is immediate transactions
covering the full range of normal customer services.

Furthermore, new shopping services can arise
with current technology. The home terminal would
permit local merchants to advertise or merely list
their merchandise on a cable TV channel. People
interested in the product could see the price and
indicate their preference to purchase some of the
items by using their two-way communicating device
in their home.

Telecommunications devices could make it possible for business executives to work in their
homes, rather than travel to an office every day.(8)

The home terminal device could provide the businessman with broadband capabilities so that he
could obtain any information he needed by dialing
it on the terminal. It could permit him to confer
with his colleagues by switched video telephone,
allow him to dictate to a secretarial pool, and
enable him to receive the finished, hard copy all
at home. Such operations are in use now, but
their cost is high. The saving in travel and
building rents may eventually reduce the costs to
nearly a trade-off with current costs of doing
business. Such a work setting would change life-
styles and habits of many people. It would require the adjustment of roles by family members
and by workers. Such a system would require a
major adjustment in business organization and
operation. It could be achieved, technologically,
and would provide the modern businessman and company with opportunities and functions far beyond
current practice.

Certainly, the information revolution, largely
the result of current telecommunications systems,
can provide the executive with a much larger amount
of information and data upon which to base decisions. The immediate analysis of vast amounts of
data is possible. The executive can request production data and plant inventories and obtain this
immediately. Then the executive can project the
possibility of meeting a particular deadline or
filling a large order. Such decisions used to be
based on hunches, conversations with plant supervisors, and expectations about the near future.
Now the decision can be based on hard data and more
precise and accurate calculations. Though not all
companies use such systems, some do, and they find
it economically beneficial. Companies which do
not use such information processing systems are or
will be placed at substantial competitive disadvantages. Many will either adopt such procedures
or lose out to competitors who have them.

Delivery of Services

Some of what has just been discussed indicates
that new communications systems will facilitate
the ways of providing services. Thus, new telecommunications devices may change established
services. For example, the way that companies
bill their customers and the methods by which they
keep their books will be changed by the inexpensive
computer linkages that are available. Many people
have had to confront a computer to deal with an
error in a bill, and they have achieved various
forms of success by "folding, spindling, or muti-
lating" their punch card. Even with these difficulties, though, all sorts of businesses find it
easier, quicker, and cheaper to bill customers by
computer services--often rented or shared computers connected to users throughout the nation
by long-distance telephone or data transmission
lines.

Education may be a major beneficiary of modern
communications systems. The use of computers to
aid instruction has developed in the past decades
because of computer interconnection permitting
students, throughout an area or even the entire
country, to tie into a single computer installation for a particular, programmed course of instruction. Despite costs, still high, computer-
aided instruction is developing and expanding into
new areas and subjects. This has not been at a
revolutionary pace, but gradual changes and developments have permitted the perfection of the
system and its expanded use.

Computers are being used, nationwide, to provide bibliographic searches. This use permits
speedy acquisition of materials on a subject without an intense effort of physically examining the
references in a library. There are specialized
services for this purpose, and there are some
general services which permit someone interested
in any one of a multitude of subjects to obtain

and use the service rapidly. Costs of a search
are currently a factor for the individual user
since they are relatively high. However, with
increased usage, especially by institutions, the
cost will decline to a point making any individual
use of the service feasible.

The delivery of health care service has been
changed and will be changed further by telecommunications devices. The continual need for doctors
to learn about recent advances in diagnosis and
treatment can and is being met by medical education systems which use cable TV channels or closed-
circuit television.

Provision of emergency medical assistance has
been facilitated by telecommunications devices
which permit the ambulance to transmit various,
vital life-sign information from the patient to
the hospital or emergency treatment center. This
can be crucial since additional minutes of monitoring and treatment can mean the difference between
life and death.

Even nonemergency transmission of medical information from one hospital to another or from the
small community to the medical center where spe-
cialists can examine and advise on the treatment
of a patient is frequent and possible because of
telecommunications systems developed in recent
years. The ATS-6 experimental satellite discussed
in Chapter 5, providing broadcast service over
India, was used for medical education and diagnosis
transmissions in Alaska, where great distances separate communities. The success of this service is
encouraging because it provides hope for better
medical service to people in remote areas.

Telecommunications systems have altered law enforcement procedures dramatically. The networking
of communications permits a local law officer to
learn in minutes if a person detained for a traffic
violation is wanted for other criminal activity by
police in other jurisdictions. Communications with

a central information center can provide the officer with nearly instantaneous information on whether a suspect is driving a stolen car. Fingerprinting, a major means of identification, can be computerized, and that process is proceeding, so that identification by such a print can occur within minutes of a request, no matter where the request comes from. Telecommunications systems have tied a number of communications centers together so that it becomes easier and quicker to deal with emergencies and information needs of a community regardless of distance or size. Centralized emergency numbers, such as "911" systems, provide a simple and direct means for any emergency to be brought to the attention of a central dispatcher who can refer the call to the appropriate agency.

New Industries

The development of technology in telecommunications has produced many far-reaching industry and economic effects. The telecommunications industry comprises a substantial portion of the industrial structure of this country as well as others. The telephone system alone requires an annual capital investment of $10 billion which is a substantial part of the nation's total private investment capital. In addition, the electronics industry, a direct result of the development of new devices and their adoption, has grown in the past decades into a billion-dollar industry which provides large amounts of materials for various operations throughout the world. The ability to miniaturize electronic circuitry has meant that various industries can be established and can operate successfully anywhere in the world. For example, the Japanese industrial development since World War II has been greatly accelerated as a result of the development of their electronics industry. Many of their materials and components are produced in this country and then shipped to Japan for assembly.

No radios or black and white television sets are
produced in the United States because it is more
economical to mass-produce these items in Japan and
ship them to the U.S. for sale.

The production of telephone equipment in this
country is divided between a major producer--Western
Electric--which produces only for the Bell system,
and small companies which make various components
for the other, independent companies. Small producers have little of the market to deal with, except that Bell apparently does contract some component production out to those companies. Western
Electric has nearly all of the "market" even
though it does not compete with anyone for the
Bell System business. A former Bell Laboratory
official feels that the production of telephone
equipment could be lost to Japanese companies if
the current Department of Justice antitrust suit
against AT&T succeeds.(9) If this happened, it
would be because the production costs for Western
Electric, reflecting actual or real costs, are
higher than for Japanese producers. Whether this
would be good for the American economy is an open
question, however. Much of the exporting of industrial work results from market factors and the
technological setting in various countries. The
ability to continue to produce this equipment in
this country will depend on continued technological advances here which reduce costs, or which
give the United States industries a technological
superiority over foreign competitors.

As telecommunications grows and changes as a
service, new companies with new functions will
arise. Whether they succeed will depend on economics and on the technological devices and service involved in the industry. Generally, the electronics industry has been quite successful in this
country although it must contend with ever-increasing foreign competition. As new breakthroughs
occur and new devices and services are adopted,

there will be incentives for continued development
of those new industries.

Life Style Changes

Much concern has been expressed about the effects
of telecommunications on individuals and on society. This is an important but also a complex subject, difficult to explore and to arrive at hard
conclusions about. Presented here are some possibilities which appear as a result of empirical
studies conducted on the subjects. There are two
dimensions to this impact--the interpersonal relations which the new devices cause to be changed,
and the individual, psychological impact of these
devices on particular persons.

Interpersonal Relations. Probably the most visible impact in this area has been the effects of mass
media on people and their lives.(10) Contact with
people is often structured around television, ad-
justing the way people think about others and the
way people react to each other. As the central
socializing agent in this country, the family has
seen its role change greatly since youngsters spend
more time before a television set than they do in
school. During his or her first 18 years a youngster
has spent an estimated 12,000 hours in the classroom
and 20,000 hours in front of a TV set.(11) The way
families are structured and the roles which parents
play have changed because of television. The parent no longer is the only source of ideas, values,
and morals for youngsters. As the technology permits, children may obtain more information and
education from a two-way communications system in
the home than from school and parents combined.
There will be a variety of societal controls over
such teaching and instruction, but the child will
also have to adapt to a changed situation in the
home, where the parent, as teacher, becomes more
separated or distant from many of the child's
experiences.

Family ties vary among families, and some ties
are as strong today as they were several generations ago. However, as the youngsters are exposed
to additional, external stimuli from mass communications systems, they will turn less to parents
for guidance and more to the other sources. In
addition, as the function of the family changes,
as it is no longer the central economic unit (as
in the case of agrarian societies), the importance
of the family bond for youngsters is likely to decline. Parents, too, will feel the pressure of
change as a result of telecommunications devices
which require an additional telephone in the home
for the child or a new room where the child can go
to learn. While this may not occur immediately,
these influences are appearing and become possible
with continuing introduction of new communications
systems and devices.

In addition to changing intrafamily relationships, changes in communications are likely to
change the setting of individuals and their relationships with fellow workers. For example, if
persons do not have to go to the "office" because
they have a terminal in their homes which gives
them access to everyone and everything they need
to perform their jobs, these persons will have to
react and adjust to this. Not having to leave the
home to work may affect not only the workers'
families, but also nonfamily members they deal with
though not having met them personally. Individuals
working in their homes may find the adjustment difficult. However, for persons who have worked at
home from the time they began elementary school, as
some future generations will, the problems will not
be the same. Generally, interpersonal relationships which result from the ease of communication
will probably affect how people react to one another. Will people find it easier or more difficult to approach strangers and strike up conversations? The audio telephone imparts a degree of
anonymity to the caller which might make it easier

for some people to use the telephone for business
and other purposes, than if they had to use a video
telephone over which they could be seen by the
other conversant.

Psychological Relations. How individuals react
to these telecommunications services will depend
largely on their psychological make-up. This, gener-
alizing about these facets is difficult. Certainly,
people have generally changed their tastes and values,
because of television and other forms of mass communication.

More individuals are being exposed to increasing
amounts of information--commercial-product information as well as decision-making information. No
matter where people live, they are exposed to the
latest fashions, the latest products, and the
latest conveniences. Whether they can obtain them
from a local store is another question, but they
at least know about them. Noncommercial information also has been presented rapidly to many
people. Thus, people face increased amounts of information about which they may know little or upon
which they may grow dependent to make decisions.
"Information overload" is a common term to describe
the phenomenon of people who face more information
than they can assimilate or cope with. This phenomenon will continue for years as communications
systems facilitate the processing and manipulation
of large amounts of information.

More people are also being exposed to universal-
ized concepts and stimuli. Thus, television viewers see the same program and the same news presentation on national networks. This may create a
commonality of information and value base which
some would argue harms the heterogeneity of the
population. Whether such common communication
channels are destroying diversity in the American
population is an interesting and an important
point. However, it is impossible to arrive at any
concrete conclusions about this possibility. The

mass communication system does provide all people
with similar or identical information about an
event. Thus, the entire country was exposed to
the difficulties and problems associated with the
Vietnam War. Each evening on the network news one
could learn what had happened in the war the same
or preceding day, regardless of where the viewer
lived. The same occurs with entertainment the networks present. Throughout the country, people view
the same drama based on the same human emergency or
emotional crisis.

Another set of impacts on individuals stem from
the socialization process which they confront as
they grow, mature, and learn from the society
around them. How a person adapts to his or her
social setting, learns the roles which he or she
will play during a lifetime, and acquires attitudes toward other people and institutions is
crucial to any society's continuing existence.

Most of us are socialized by our parents, our
peers (fellow students, workers, and associates),
and by general sources of attitudes such as mass
media. Traditionally the socialization process
has largely resulted from parental guidance and
schooling which occurs during a youngster's early
years. However, research suggests that socialization continues throughout life, and the school,
peers, and media appear to have a greater influence on socialization than do parents. Depending
on the parents and their willingness and abilities,
a youngster can gain nearly all attitudes and
orientations from peers and teachers without registering much parental impact at all. As a result,
telecommunications, particularly mass media, can
play a major role in the socialization process of
youngsters. It is this that has caused much concern about the generation gap which some observers
see developing.

While the socialization process is important and
should be considered in terms of telecommunications

technology, it is also important to consider the
fact that much of this impact on people may well
be unintended. Some observers argue that the mass
media are consciously trying to develop certain
kinds of attitudes and biases among the population.(2,12) There is certainly much concern about
this and probably some truth to it. However, *even
if* one assumes that the media are well-intentioned,
there are many indications that programming and
presentations give people particular perspectives,
which they in turn act upon or at least use as a
basis for their orientations towards others. In
some situations, a person has learned to respond
to situations from having watched a television program in which the identical situation arose. In
other situations, the individual has adopted an
entire identity and self-image which shapes his
personality from media sources. Whether this
clearly results from media programming, intended
or unintended, or from other influences, the impact seems clear and should be considered by those
interested in learning how the technology will affect the individual.

A major area of controversy about the impact of
mass communications on society has been the effect
of violence and violence-oriented television programs on children who watch them. Basic conclusions of researchers in this area do not *prove*
that violent programs make children violent, but
suggests that this can, and does, occur. Generally, a youngster with aggressive tendencies, or
so inclined by other stimuli, is more likely to
act out aggressions if they are reinforced by programming viewed on television. Competing theories
include the possibility that television violence
acts as a catharsis and replaces acting out of aggressive tendencies among watchers. However, there
is good support for the proposition that some viewers, with aggressive tendencies, gain reinforcement
for these from violent television programs, and may

act on these feelings as a result of viewing programs.(10,13,14,15)

Obviously, this area is important for society
and various suggestions to reduce the amount of
violent TV programming have been made. However,
control of programs apparently will not come from
voluntary network efforts, because violent programs seem to attract the most viewers and thus
generate the greatest advertising revenue. Any
governmental policy limiting the amount of violent
programming would constitute prior censorship and
would have to face a substantial court test regarding its constitutionality. The solution to
this dilemma is unclear, and the final outcome may
well be slight adjustments in the status quo which
do not satisfy either side completely, but which
may compromise their differences to the point of
acceptance. However, the basic point is that programming can, and probably does, affect viewers,
and, given the large amount of exposure to television which many children have, this effect is a
serious social problem.

Another set of communication system problems relates to the individual psychological make-up of
persons. This involves the anonymity which rapid
and profuse communication systems provide. The
individual becomes a series of numbers ranging from
a social security number to charge account numbers
issued by various commercial establishments. Some
people will find this impersonalization destructive
and disturbing. Coupled with the increased lack of
privacy which can arise in a highly developed communications infrastructure, this phenomenon presents great problems for individuals and for society as a whole.

Many devices are operational and permit various
sorts of data collection, ranging from tracking or
locating devices to wiretapping and eavesdropping
mechanisms. In addition, possible future devices
that permit paging and telephoning systems to

reach anyone, anywhere, could create problems for
individuals. The current growing concern about invasions of privacy presents important problems for
society, and, as the devices for such invasions become more sophisticated, the problems will magnify.
The thought of not being able to be "alone" causes
deep concern for many people who view this as a
transgression on a basic human, if not constitutional, right. For some this possibility would
destroy a person's individuality and the ability
to maintain his or her identity.

The Impacts

These suggestions of technological impact have only
touched selected future problems. Some of the
problems may never occur unless certain technologies and devices develop and are used widely.
Thus, the broadband home terminal system will not
require major adjustments in work styles or family
and educational pattern if the systems are not
well-developed and widely-distributed to the general population or if they emerge only slowly,
allowing time for adjustment and adaptation. One
development may have one kind of impact on American
society, because of its highly developed social
system, but may have a much different impact on
another culture or society at a different stage of
development. Thus, some impacts may be long range
and depend on the speed of development as well as
the target government or society.

Other impacts, clearly with us, present major
problems for society. The telephone has been with
us for many years, yet new telephone services are
being offered almost daily, and these services
require adjustments in life styles and commercial
transactions as well as changes in political processes. Television is a widespread phenomenon in
America, and, even if the precise effect of television is unclear, many examples and evidence suggest some of the major impacts which seem to result

from this telecommunication system. The development and growth of industries, even national industrial bases, are realities which are largely due to telecommunications technology.

Generally, the increase in communications and the ability to communicate rapidly with increasingly larger numbers of people is probably the major story of the technological impact. How our society or any other copes with this many-faceted set of pressures and problems is a major question. It is clear that systematically, or haphazardly, societies must cope with this ever-increasing and widespread net of problems.

Chapter 7

Future Telecommunications Developments

This book has explored how telecommunications technology has developed in the United States. To determine why this development occurred as it did,
this book has examined the interactions of the
forces among various political and commercial institutions. This chapter will suggest structural or
policy changes, changes that might be made, which
might alter constructively future developments of
telecommunications technology. None of these suggestions is provided as the answer to a specific
problem, and most of them carry with them or create
some problems of their own. Hopefully, however,
this discussion will provide the reader with
thought-provoking material about what has happened
and what ought to happen.

Alternatives to Present Regulatory Approaches

Telecommunication services have been expanding and
improving continuously since their inception more
than 100 years ago. Case studies in Chapters 4 and
5 show many forces at work in determining when and
how various telecommunication services are provided. In some cases, potential services may
never develop because political constraints are
only marginally related to the public interest.
In other cases, political and regulatory agencies
may attempt to stimulate development of new serv-

ices by requiring that certain technological capabilities be provided or by permitting competitive
situations to develop in an otherwise monopolistic
industry. The two methods which appear to have
been relatively the most effective in controlling
new services are: the withholding of authorization
for new entrants or the promulgation of restrictive
and stifling regulations for such entrants; and
opening up an area to free-market competition.
Other procedures have been much less effective.
In general most of the regulatory agency decisions
are made so they do not adversely affect existing
services of established industries. This means
that the pivotal decisions are most likely to involve new services. Therefore, in this area consideration will be given to alternative approaches
aimed at introducing a different balance of forces
in the decision process.

   With a change in the balance of forces it appears possible to alter substantially the evolution of telecommunications development. The key
factor in any such procedure is a feedback mechanism that will prevent continuation of the development and evolution process when such continuation is not in the public interest. This factor
is provided by competitive forces in a free market
and is lacking in a monopolistic industry. Introduction of an equivalent effect can be accomplished
by other means, and some such possibilities will be
considered next.

Pay Television

An example of restrictive regulations is the current status of pay television technology and service. National networks and the TV broadcasting
industry oppose pay television, seeing it as a
major threat to the size of their audiences, which,
in turn, would be a threat to advertising revenues.
The broadcasters' strongest argument that relates
to public interest is that the advent of pay TV

would eliminate "free television" and would represent a major hardship on the poor. This argument is difficult to overcome without changing some rules of operation for pay TV from those being considered. There seems little doubt that the main function of TV from the advertiser's and broadcaster's point of view is not entertaining the poor but, rather, influencing the purchasing habits of the nonpoor. The free entertainment of the poor is a fringe benefit to society paid for by the more affluent who purchase products advertised on TV.

For the sake of argument, assume that the existence of free television programming and pay television and its potentially greater range of offerings are both desirable. How might these two objectives be made mutually achievable? One approach might be to make pay TV available only at nonprime time--say from 10 p.m. to 1 a.m. A similar result might be accomplished by limiting pay TV to certain specified nights such as Monday and/or Thursday. These arrangements would provide a substantial audience for advertiser-sponsored programming while permitting major pay TV productions an opportunity to gain sufficient viewers to make them profitable. Some additional requirements might provide sufficient competition to protect viewer interests. For example, pay TV producers might be required to provide a specified amount of nonpay TV programming that would be generally available. Furthermore, attainment of a specified fraction of the potential viewing audience of the free programming might be required in order that the pay TV broadcaster retain authority to offer programs. The second requirement would assure that the quality of the free programming remain high. If it did not, that is, if viewer surveys indicated that people did not watch it, then the pay TV producer would have strong incentive to improve his programming because of the threat to his license. This is an example of consumer feedback.

Free programming requirements could be changed
in terms of time or audience penetration as a
means of controlling the availability and quality
of such programming. Free programming would compete with advertiser-sponsored programming or it
might be advertiser-sponsored itself. The key
factor is sampling users of free programming to
measure its quality. The marketplace would set
the level of quality of pay TV programming.

Telephone System

Because of the vast resources of capital, technology, manufacturing, and marketing, it is unrealistic to expect any meaningful competition in
primary services supplied by the Bell System. The
telephone industry has operated as a regulated
monopoly for decades. By almost any standard of
comparison, that operation has been highly successful and has led to an outstanding telephone service
in this country--unquestionably the best in the
world. Nevertheless, even though progress in
telephone services has been continuous and substantial, it is not obvious that their development
has been optimal.

In many ways the telephone companies operate
as a benevolent monopoly. Meeting their customer
needs is based on their judgment and evaluation of
those needs and their decision as to whether those
needs are being met appropriately at any particular
time. Decisions to provide specific new services,
for example, automatic fire alarms or automatic
telephone answering, appear to be made with little
weight given to consumer preferences other than
whether such services would be marketable. Since
there are more technologically feasible and marketable services than there are capital resources to
provide them, the decision generally is not heavily
weighted in terms of consumer preferences. Because of the monopolistic nature of the telephone
system, the consumer has virtually no recourse if

he does not agree with the policies being pursued by the telephone company. The ineffectiveness of the regulatory action in controlling services is clearly evidenced by the great disparity between services different telephone companies supply to adjacent communities. A frequent complaint of non-Bell customers is that they have outdated and inadequate telephone services compared with what is already in service in a large part of the Bell System.

The question that must be addressed is whether or not it is possible to introduce into the telephone system a new force that will provide some of the desirable features of open competition in providing an input for customer preferences. FCC attempts to introduce competition have not been successful. Either the telephone company has chosen not to compete or to compete so strongly that it virtually overpowers the new entrant.

An example of noncompetition by the telephone company is the *Carterfone* decision discussed in Chapter 5. In this case the FCC ruled that apparatus not owned by the telephone company could be connected into the telephone system. After some attempt to have the ruling overturned in the courts, AT&T dropped its formal opposition. In some areas such as PBX exchanges, AT&T is attempting to preserve a competitive position but in most areas of foreign attachments the telephone company is making no attempt to compete. Thus, an essentially noncompetitive situation exists as far as the telephone company is concerned. One thing this means is that the vast engineering and manufacturing resources of AT&T will not be brought to bear on this area of technological development.

An example of the opposite reaction by the telephone company is the *MCI* decision also discussed in Chapter 5. In this case the FCC ruled that Microwave Communications, Inc., would be licensed to provide certain intercity data transmission

services not then available from the telephone
company. Again, as in the *Carterfone* case, AT&T
resisted this decision through court action. More
importantly, however, AT&T undertook an acceler-
ated research and development program to improve
the range and quality of data transmission services
that it could supply. Its development program was
highly successful and it was able to supply a wide
range of data transmission services at low costs--
lower, in fact, than those for which MCI could
provide the same services. There seems little
doubt that the same, or equivalent, data transmission capabilities would eventually have been
developed on a timetable established by the telephone company.

This particular intervention by the FCC altered
that timetable and caused a reshuffling of substantial engineering and other resources by the
Bell system. Whether, in the long run, this forced
competition resulted in a net improvement of the
telecommunications system or whether it just inter-
changed the timetable of certain of the services
will be difficult to determine. However, it seems
unlikely that the crash development of the data
transmission capability is likely to have been as
efficiently conducted as if it had been part of a
more carefully planned program.

On the other hand, without the incentive introduced by competition from a new entrant, the re-
ordering of priorities would not have occurred and
data transmission costs would not have undergone
the dramatic decrease that they did at that particular time. At best, the introduction of direct
competition into markets dominated by the telephone companies appears to be of doubtful value.
Some other technique must be employed to obtain
the benefits equivalent to those resulting from
competition without attempting to destroy the
natural monopoly characteristics of telephony
itself.

Synthetic Competition. In a free enterprise
system competition among various corporations
vying for a share of the market is expected. Variations in price, product or service characteristics,
and marketing strategy will lead to preferential
selection of customers. The product or service
finding greatest acceptance with the customers
will produce an imbalance in sales that will cause
companies that are noncompetitive to alter their
approach in such a manner as to make their product
or service more attractive to prospective customers. Many situations arise that prevent the ideal
operation of a competitive free enterprise system,
and the automatic correction caused by feedback
through customer selection may be absent. Such a
situation occurs when a monopoly exists and a single company controls the diversity of products and
services. This problem is particularly aggravated
when the monopolist has sufficiently large resources to suppress competition by undercutting
prices in areas where such competition is attempted. Under such conditions no significant
competition can result and, therefore, customer
preferences are given only minor weight in determining policies.

Recently, the FCC instituted a fact-finding
proceeding with regard to the role of competition
in the telecommunications industry.(15) With regard to this inquiry, AT&T offered the following
comments(16):

1. Telecommunications services are most widely
available at the lowest overall cost to the entire
public and the country when those services are provided by regulated monopolies.

2. When competition is introduced in the telecommunications industry, it results in higher overall
costs and in increased rates to exchange service
users. These adverse effects on exchange users
can be mitigated if existing carriers are allowed
to compete on an equal basis.

3. The public interest will be most adversely af-
fected--in terms of service availability, cost and
quality--when telecommunications services are provided under a system of contrived competition or
arbitrary market allocation.

These conclusions stem from the fact that the
telecommunications industry displays economies of
scale and the characteristics of a natural monopoly in the provision of services as a whole.

Theory suggests that since natural monopoly
characteristics are present, the telecommunications field should hold little attraction for competitors. That would be true today if the industry
had not followed pricing policies explicitly designed to attain the goal of universal service.
This goal was made the law of the land in the Communications Act of 1934:

"...to make available so far as possible, to all
the people of the United States, a rapid, efficient,
Nation-wide and world-wide wire and radio communications service with adequate facilities at reasonable charges..."

Ironically the very success of the regulated
telecommunications industry in reaching the goal
of universal service at reasonable rates makes it
vulnerable to selective competition, because this
success has been achieved in great part through
the application of value-of-service pricing and
nationwide and statewide rate and cost averaging.
This type of pricing, approved and even encouraged
by regulators at both the state and federal levels,
invites selective entry into those parts of the
industry that can be served with the least cost
and at the highest profit. Yet these are the very
pricing policies that give the service its widest
availability.

It appears that the telephone industry can make
a very strong case to support its monopolistic position and it seems improbable that it will be

seriously threatened in the near future. Some
alternative approach would seem indicated.

The FCC recently issued its decision regarding
this major inquiry and found very little to criti-
cize about the AT&T rate base or methods of operation. The agency did ask that AT&T justify some
of its practice more precisely than had been done
to date. However, the thrust of the commission's
decision was that the services and the prices
charged were excellent and reasonable, respectively.
The arguments and evidence presented by the common
carrier were apparently quite compelling when the
commission gave them close scrutiny.

In situations where the possibility of free
market competition does not exist, it would appear
reasonable to attempt to achieve some of the desirable features of such competition by other means.
Several possibilities appear to have sufficient
promise to warrant further consideration. Two of
the highly desirable primary benefits of competition are diversity of offerings allowing customer
preference to be exhibited and pressure to keep
prices low to increase the fraction of the market
a particular company can capture. Because of the
regulated nature of telecommunications, it may not
be necessary to provide for price competition but
rather to rely on price justification as is done
now. If this is so, then the primary requirement
would be some type of product and service diversity that would provide accurate customer preference feedback.

One method of generating customer-preference
feedback in the telephone industry would be to require that a substantially increased range of
services and equipments be made available to customers. These need not be made available to all
users but might be provided in selected test or
control regions. The results of such experiments
would be required inputs to the regulatory agency
for the approval of new or the adjustment of old

services. In this way customers would have a way
of registering their preferences meaningfully--
comparable in many ways to a competitive situation.
Some thought would have to be given to the ways in
which the various alternative services and devices
would be selected. One approach might be the joint
agency/industry development of a list of potential
services and devices which must be made available
to test groups of customers before such services
are inaugurated in any sizeable way.

In keeping with any attempt to have the regulatory agency engage more actively in determining
what technology will be developed, apparently some
restructuring of the FCC should be done. Today the
FCC consists of commissioners with legal back-
grounds. The introduction of stronger technological and consumer-oriented voices on the commission
would provide a more balanced approach to the problems inherent in any attempt to provide a simulated
competitive situation for determining the directions of technological evolution.

Selective Competition. Another approach to providing competition-induced feedback of customers
is to permit open competition in certain areas.
Examples of this are foreign attachments to the
telephone system, the supplying of private leased
lines by special carriers, and paging services.
In some developing areas competition also appears
feasible. The most imminent is that of mobile
telephone services. They could well be provided
competitively within a single geographical area
without any danger of damaging the integrity of
the switched telephone system.

Other possibilities for direct competition might
be developed from the lists of potential technological services and devices that are prepared for the
simulated competition approach. Some of the proposed devices or services might be of no interest
to the telephone company and, therefore, could be
made available through open competition. In the

long run it may be that supplying the distribution, switching, and transmission of signals becomes the primary function of the telephone company
while generation and use of the signals is the primary function of broad-based industrial competitive
activities. If such services included the handling
of wideband signals such as those of CATV and the
full ramifications of the wired city, the magnitude of the physical plant and the capital investment involved would be of such proportions as to
dwarf present telephone company investments.

Data Transmission

The transmission of information other than conventional radio, television, and voice signals is
expanding at an enormous rate. This category of
telecommunications called data transmission includes such things as airline reservations; accounting, inventory, and production control operations of companies; linking of computers at widely
separated locations; and interrogation and response
of remote data banks, ranging from credit ratings
to criminal records. These will eventually include
many presently undreamed of services. As we move
toward a cashless society and an electronic postal
service this kind of information will far exceed
that handled by the conventional switched telephone system. Meanwhile, there is no way that
data transmission systems can operate completely
independently of the switched telephone system.
This is because isolated users, such as retail
outlets or individuals in their homes, need to be
able to access the data transmission system in an
economical manner, and this implies use of the
already available telephone system.

  The direction of future developments in this
area is complex. Already, several highly di-
vergent approaches are being employed in the development of data transmission systems. Common
carriers, such as AT&T, have microwave and coaxial

cable systems suitable for both voice and data
transmission; the specialized carriers such as MCI
have dedicated microwave systems for data transmission; and a whole host of companies are developing
satellite systems with multiple point-to-point
data transmission capabilities.

AT&T takes the position that denying it the opportunity to offer data transmission services will
unquestionably lead to substantially higher telephone rates due to loss of subsidies that data
transmission services provide. Yet, it is clear
that meaningful competition with common carriers
in a situation where they have a guaranteed use
of voice services is not possible. It would become impossible to tell which service was subsi-
dizing the other.

Because of the anticipated growth in data transmission via satellites the inherent advantages that
the common carriers have in already-installed
ground-based microwave systems will no longer be
very important. The data transmission requirements
will overshadow the telephone usage in a few years
and will be the dominant factor in the competitive
situation. In the long run it appears that free
competition for data transmission services via
satellite may be a feasible approach to control of
the marketplace. In such a case participation by
the common carriers would be on an equal basis
with other companies and they would not enjoy any
special advantage because of their current responsibilities of providing low cost telephone services. It would be desirable for the large common
carriers such as AT&T and GTE to enter this market
because of the strong contributions their technical
expertise would make to both quality and cost of
service. However, their entry would be as an equal
competitor rather than with any substantial advantage due to their telephone operation.

In the future, data transmission services will
be as important a part of a nation's life blood

as the telephone is today. The same kind of precautions will have to be taken to assure continuity
and reliability of service as is done now with the
telephone. A major factor in such precautions may
well be having available multiple, competing services that can interchange transmission capacity
among each other. Many services such as banking,
criminal data banks, military logistics, or air
traffic control will require extraordinary reliability to avoid the disastrous results which could
occur with loss of such services for even a short
period of time. Other types of services of a more
routine nature would be only slightly affected by
short down-times for the systems. Careful planning
will be required to insure intersystem compatibility and interconnectability. Regulatory and
legal decisions made now will have lasting effects
on how the data transmission services of the future develop and where the line is drawn between
the natural monopoly of the switched telephone
system and competitive data transmission services.

Conclusions

The development of telecommunications technology
and its uses are strongly influenced by many
factors discussed in earlier sections of this
book. The rate of development and use, as well as
the direction of evolution, depend on a wide variety of political forces and processes, in addition
to economic variables which structure such development. In a major way these nontechnological considerations impinge on and control telecommunications development and growth. The results of this
complex interplay depend upon which forces or
factors dominate a particular issue or policy process that is crucial to the outcome. The FCC policies on cable TV illustrate the traditional
pattern of protecting the status quo, while seeking to achieve an acceptable compromise with the

challenging interest on this issue. While this
outcome is often the pattern that emerges from
the political allocation process, it is not always
the end result, as in the cases of *Carterfone* and
MCI efforts to produce competition with the large
common carriers. Certainly, the outcome is not
pre-ordained in telecommunications issues.

Two general categories of efforts have been
tried to achieve changes in the development of
technology. The first involves the development
of new entrants and challengers who seek to dislodge the established industry in a head-on con-
frontation in the political arenas. Challengers
have tried this often, but with little success.
Usually, the potent political power of the established groups dominates in such an encounter. The
second category of efforts relates to changing
either the institutions which make the allocations
or changing the processes by which the allocative
decisions are made.

What changes made in the processes or institutions seek to accomplish is to permit access to
new groups of interests (e.g., consumers). Or they
give new or different forces the ability to participate in decision-making, or even to control or
dominate such decision-making. The success of
such adjustments is never certain beforehand.
However, given the existing processes and their
results, certain changes in the FCC or in the
other allocative processes could be made with fair
certainty that particular problems would be eliminated. The objective of such institutional changes,
however, is not just to alter structure. Rather,
the purpose is to achieve different policies regarding the use and development of technology.

The discussions in this book illustrate the
complexity of the forces and constraints operating
to control developments in telecommunications.
These forces and constraints are only partially
understood and are themselves constantly changing

so that the process is dynamic from both the
technological and the political point of view.
A much better understanding of these processes
will be required before any planned technological
evolution can hope to succeed. Some FCC attempts
to stimulate development in certain areas or to
introduce competition indicate an awareness of the
need for altering or influencing technological developments. However, until there is better understanding of the social, political, and technological
factors in such processes there is little hope that
they will accomplish their objectives.

An alternative procedure that has some promise
is to cause innovations to be introduced in a
manner that will provide feedback from users as
well as suppliers so as to provide a self-
stabilizing marketplace in the absence of price
competition. Such a procedure permits acquisition
of data based on actual user experience with various telecommunications services before large scale
commitments of resources are made.

In some situations such as data transmission
by satellite and dedicated surface transmission
links it appears feasible to permit a competitive
situation to develop. Where such data transmission operations provide services vital to the day-
to-day welfare and safety of the public, it will
be necessary to provide safeguards against cata-
strophic loss of services by requiring standardization of transmission formats and rerouting
capabilities so that loss of some part of a system will not mean loss of data transmission capability.

Appendix

The purpose of this appendix is to present some of the exact
wording which appears in various statutes, federal regulations,
and court and FCC decisions relating to the regulation of
telecommunications. This should give the interested reader
an opportunity to see exactly what the "words" provide for in
the various documents. Clearly, the words are not the only
important dimensions of these documents, since their impacts
reach far beyond the words. The interpretations placed on
the documents by courts, the commission, and others clearly
outline more fully than just the words what effect they will
have.

Several explanatory notes should be provided here. Particular portions of these documents provide the basis for
particular policies which have been discussed earlier in this
book. For example, the All-Channel Receiver law is contained
in § 303(s) of Title 47 of the *United States Code*. This brief
paragraph forms the statutory authority for a small portion of
the *Code of Federal Regulations* (§§ 15.65-15.68) which may
have a profound effect on the development of UHF television
broadcasting. The language of the Supreme Court in *Southwestern Cable*, while vague, provides the legal authority for
all of § 76 of the *Code of Federal Regulations* dealing with
cable television regulations, a small portion of which is reprinted in this appendix. The Communications Satellite Act
of 1962 is contained in 47 U.S.C. §§ 701 ff. Although only
portions of it are reprinted, they provide the essence of the
statute creating an entire structure for development of the
communications satellite industry. Where appropriate, the
statutory wording is provided, and then either the regulation
or court decision wording pertaining to particular parts of
the statute are provided. In this way the reader can see how
the commission or court has used the law, to develop a pol-
icy--at least in terms of wording.

The Communications Act of 1934 as Amended.

This statute is found in Title 47 of the *United States Code*
(47 U.S.C.). Particular sections of this statute are presented here to give the reader a picture of the words authorizing the FCC to regulate wire and radio communications in
this country.

§ 151. *Purposes of chapter; Federal Communications Commission
        created*
For the purpose of regulating interstate and foreign commerce
in communication by wire and radio so as to make available,
so far as possible, to all the people of the United States a
rapid, efficient, Nation-wide, and world-wide wire and radio
communication service with adequate facilities at reasonable
charges, for the purpose of the national defense, for the purpose of promoting safety of life and property through the use
of wire and radio communication, and for the purpose of secur-
ing a more effective execution of this policy by centralizing
authority heretofore granted by law to several agencies and by
granting additional authority with respect to interstate and
foreign commerce in wire and radio communication, there is
created a commission to be known as the "Federal Communications
Commission", which shall be constituted as hereinafter provided, and which shall execute and enforce the provisions of
this chapter.

                SUBCHAPTER II. COMMON CARRIERS
§ 201. *Service and charges*
    (a) It shall be the duty of every common carrier engaged
in interstate or foreign communication by wire or radio to
furnish such communication service upon reasonable request
therefor; and, in accordance with the orders of the Commission,
in cases where the Commission, after opportunity for hearing,
finds such action necessary or desirable in the public interest, to establish physical connections with other carriers, to
establish through routes and charges applicable thereto and
the divisions of such charges, and to establish and provide
facilities and regulations for operating such through routes.
    (b) All charges, practices, classifications, and regula-
tions for and in connection with such communication service,
shall be just and reasonable, and any such charge, practice,
classification, or regulation that is unjust or unreasonable
is declared to be unlawful: *Provided,* That communications by
wire or radio subject to this chapter may be classified into
day, night, repeated, unrepeated, letter, commercial, press,
Government, and such other classes as the Commission may decide to be just and reasonable, and different charges may be
made for the different classes of communications: *Provided
further,* That nothing in this chapter or in any other provision of law shall be construed to prevent a common carrier

subject to this chapter from entering into or operating under
any contract with any common carrier not subject to this chapter, for the exchange of their services, if the Commission is
of the opinion that such contract is not contrary to the public interest:

* * *

§ 205.   *Commission authorized to prescribe just and reasonable*
         charges; penalties for violations
   (a)   Whenever, after full opportunity for hearing, upon a
complaint or under an order for investigation and hearing made
by the Commission on its own initiative, the Commission shall
be of opinion that any charge, classification, regulation, or
practice of any carrier or carriers is or will be in violation
of any of the provisions of this chapter, the Commission is
authorized and empowered to determine and prescribe what will
be the just and reasonable charge or the maximum or minimum,
or maximum and minimum, charge or charges to be thereafter ob-
served, and what classification, regulation, or practice is or
will be just, fair, and reasonable, to be thereafter followed,
and to make an order that the carrier or carriers shall cease
and desist from such violation to the extent that the Commission finds that the same does or will exist, and shall not
thereafter publish, demand, or collect any charge other than
the charge so prescribed, or in excess of the maximum or less
than the minimum so prescribed, as the case may be, and shall
adopt the classification and shall conform to and observe the
regulation or practice so prescribed.
   (b)   Any carrier, any officer, representative, or agent of
a carrier, or any receiver, trustee, lessee, or agent of either
of them, who knowingly fails or neglects to obey any order made
under the provisions of this section shall forfeit to the
United States the sum of $1,000 for each offense. Every dis-
tinct violation shall be a separate offense, and in case of
continuing violation each day shall be deemed a separate offense.

§ 301.   *License for radio communication or transmission of*
         energy
It is the purpose of this chapter, among other things, to maintain the control of the United States over all the channels of
interstate and foreign radio transmission; and to provide for
the use of such channels, but not the ownership thereof, by
persons for limited periods of time, under licenses granted by
Federal authority, and no such license shall be construed to
create any right, beyond the terms, conditions, and periods of
the license. No person shall use or operate any apparatus for
the transmission of energy or communications or signals by radio (a) from one place in any Territory or possession of the
United States or in the District of Columbia to another place

in the same Territory, possession, or District; or (b) from
any State, Territory, or possession of the United States, or
from the District of Columbia to any other State, Territory,
or possession of the United States; or (c) from any place in
any State, Territory, or possession of the United States, or
in the District of Columbia, to any place in any foreign
country or to any vessel; or (d) within any State when the
effects of such use extend beyond the borders of said State,
or when interference is caused by such use or operation with
the transmission of such energy, communications, or signals
from within said State to any place beyond its borders, or
from any place beyond its borders to any place within said
State, or with the transmission or reception of such energy,
communications, or signals from and/or to places beyond the
borders of said State; or (e) upon any vessel or aircraft of
the United States; or (f) upon any other mobile stations within the jurisdiction of the United States, except under and in
accordance with this chapter and with a license in that behalf
granted under the provisions of this chapter.

§ 303. *Powers and duties of Commission*
Except as otherwise provided in this chapter, the Commission
from time to time, as public convenience, interest, or necessity requires, shall--

   (a) Classify radio stations;
   (b) Prescribe the nature of the service to be rendered by
each class of licensed stations and each station within any
class;
   (c) Assign bands of frequencies to the various classes of
stations, and assign frequencies for each individual station
and determine the power which each station shall use and the
time during which it may operate;
   (d) Determine the location of classes of stations or individual stations;
   (e) Regulate the kind of apparatus to be used with respect
to its external effects and the purity and sharpness of the
emissions from each station and from the apparatus therein;
   (f) Make such regulations not inconsistent with law as it
may deem necessary to prevent interference between stations
and to carry out the provisions of this chapter: *Provided, however,* That changes in the frequencies, authorized power, or in
the times of operation of any station, shall not be made without the consent of the station licensee unless, after a public
hearing, the Commission shall determine that such changes will
promote public convenience or interest or will serve public
necessity, or the provisions of this chapter will be more fully
complied with;
   (g) Study new uses for radio, provide for experimental
uses of frequencies, and generally encourage the larger and
more effective use of radio in the public interest;

(h) Have authority to establish areas or zones to be served by any station;

(i) Have authority to make special regulations applicable to radio stations engaged in chain broadcasting;

(j) Have authority to make general rules and regulations requiring stations to keep such records of programs, transmissions of energy, communications, or signals as it may deem desirable;

(k) Have authority to exclude from the requirements of any regulations in whole or in part any radio station upon railroad rolling stock, or to modify such regulations in its discretion;

(l) Have authority to prescribe the qualifications of station operators, to classify them according to the duties to be performed, to fix the forms of such licenses, and to issue them to such citizens or nationals of the United States as the Commission finds qualified, except that in issuing licenses for the operation of radio stations on aircraft the Commission may, if it finds that the public interest will be served thereby, waive the requirement of citizenship in the case of persons holding United States pilot certificates or in the case of persons holding foreign aircraft pilot certificates which are valid in the United States on the basis of reciprocal agreements entered into with foreign governments;

(m)(1) Have authority to suspend the license of any operator upon proof sufficient to satisfy the Commission that the licensee--

(A) has violated any provision of any Act, treaty, or convention binding on the United States, which the Commission is authorized to administer, or any such Act, treaty, or convention; or

(B) has failed to carry out a lawful order of the master or person lawfully in charge of the ship or aircraft on which he is employed; or

(C) has willfully damaged or permitted radio apparatus or installations to be damaged; or

(D) has transmitted superfluous radio communications or signals or communications containing profane or obscene words, language, or meaning, or has knowingly transmitted--

(1) false or deceptive signals or communications, or

(2) a call signal or letter which has not been assigned by proper authority to the station he is operating; or

(E) has willfully or maliciously interfered with any other radio communications or signals; or

(F) has obtained or attempted to obtain, or has assisted another to obtain or attempt to obtain, an operator's license by fraudulent means.

(2) No order or suspension of any operator's license shall take effect until fifteen days' notice in writing thereof, stating the cause for the proposed suspension, has been given to the operator licensee who may make written

application to the Commission at any time within said fifteen days for a hearing upon such order. The notice to the operator licensee shall not be effective until actually received by him, and from that time he shall have fifteen days in which to mail the said application. In the event that physical conditions prevent mailing of the application at the expiration of the fifteen-day period, the application shall then be mailed as soon as possible thereafter, accompanied by a satisfactory explanation of the delay. Upon receipt by the Commission of such application for hearing, said order of suspension shall be held in abeyance until the conclusion of the hearing which shall be conducted under such rules as the Commission may prescribe. Upon the conclusion of said hearing the Commission may affirm, modify, or revoke said order of suspension.

(n) Have authority to inspect all radio installations associated with stations required to be licensed by any Act or which are subject to the provisions of any Act, treaty, or convention binding on the United States, to ascertain whether in construction, installation, and operation they conform to the requirements of the rules and regulations of the Commission, the provisions of any Act, the terms of any treaty or convention binding on the United States, and the conditions of the license or other instrument of authorization under which they are constructed, installed, or operated.

(o) Have authority to designate call letters of all stations;

(p) Have authority to cause to be published such call letters and such other announcements and data as in the judgment of the Commission may be required for the efficient operation of radio stations subject to the jurisdiction of the United States and for the proper enforcement of this chapter;

(q) Have authority to require the painting and/or illumination of radio towers if and when in its judgment such towers constitute, or there is a reasonable possibility that they may constitute, a menace to air navigation.

(r) Make such rules and regulations and prescribe such restrictions and conditions, not inconsistent with law, as may be necessary to carry out the provisions of this chapter, or any international radio or wire communications treaty or convention, or regulations annexed thereto, including any treaty or convention insofar as it relates to the use of radio, to which the United States is or may hereafter become a party.

(s) Have authority to require that apparatus designed to receive television pictures broadcast simultaneously with sound be capable of adequately receiving all frequencies allocated by the Commission to television broadcasting when such apparatus is shipped in interstate commerce, or is imported from any foreign country into the United States, for sale or resale to the public.

§ 701. *Congressional declaration of policy and purpose*
    (a) The Congress declares that it is the policy of the
United States to establish, in conjunction and in cooperation
with other countries, as expeditiously as practicable a commercial communications satellite system, as part of an improved
global communications network, which will be responsive to public needs and national objectives, which will serve the communication needs of the United States and other countries, and
which will contribute to world peace and understanding.
    (b) The new and expanded telecommunication services are to
be made available as promptly as possible and are to be extended to provide global coverage at the earliest practicable date.
In effectuating this program, care and attention will be directed toward providing such services to economically less developed countries and areas as well as those more highly developed,
toward efficient and economical use of the electromagnetic frequency spectrum, and toward the reflection of the benefits of
this new technology in both quality of services and charges for
such services.
    (c) In order to facilitate this development and to provide
for the widest possible participation by private enterprise,
United States participation in the global system shall be in
the form of a private corporation, subject to appropriate governmental regulation. It is the intent of Congress that all
authorized users shall have nondiscriminatory access to the
system; that maximum competition be maintained in the provision
of equipment and services utilized by the system; that the
corporation created under this chapter be so organized and
operated as to maintain and strengthen competition in the provision of communications services to the public; and that the
activities of the corporation created under this chapter and
of the persons or companies participating in the ownership of
the corporation shall be consistent with the Federal antitrust
laws.
    (d) It is not the intent of Congress by this chapter to
preclude the use of the communications satellite system for
domestic communications services where consistent with the provisions of this chapter nor to preclude the creation of additional communications satellite systems, if required to meet
unique governmental needs or if otherwise required in the national interest.

§ 721. *Implementation of policy*
In order to achieve the objectives and to carry out the purposes of this chapter--
    (a) the President shall--
    (1) aid in the planning and development and foster the
execution of a national program for the establishment and
operation, as expeditiously as possible, of a commercial
communications satellite system;

(2) provide for continuous review of all phases of the
development and operation of such a system, including the
activities of a communications satellite corporation authorized under subchapter III of this chapter;

(3) coordinate the activities of governmental agencies
with responsibilities in the field of telecommunication, so
as to insure that there is full and effective compliance at
all times with the policies set forth in this chapter;

(4) exercise such supervision over relationships of the
corporation with foreign governments or entities or with
international bodies as may be appropriate to assure that
such relationships shall be consistent with the national interest and foreign policy of the United States;

(5) insure that timely arrangements are made under which
there can be foreign participation in the establishment and
use of a communications satellite system;

(6) take all necessary steps to insure the availability
and appropriate utilization of the communications satellite
system for general governmental purposes except where a
separate communications satellite system is required to
meet unique governmental needs, or is otherwise required
in the national interest; and

(7) so exercise his authority as to help attain coordinated and efficient use of the electromagnetic spectrum and
the technical compatibility of the system with existing
communications facilities both in the United States and
abroad.

(b)   the National Aeronautics and Space Administration
shall--

(1) advise the Commission on technical characteristics
of the communications satellite system;

(2) cooperate with the corporation in research and development to the extent deemed appropriate by the Administration in the public interest;

(3) assist the corporation in the conduct of its research
and development program by furnishing to the corporation,
when requested, on a reimbursable basis, such satellite
launching and associated services as the Administration
deems necessary for the most expeditious and economical development of the communications satellite system;

(4) consult with the corporation with respect to the
technical characteristics of the communications satellite
system;

(5) furnish to the corporation, on request and on a reimbursable basis, satellite launching and associated services required for the establishment, operation, and maintenance of the communications satellite system approved by
the Commission; and

(6) to the extent feasible, furnish other services, on
a reimbursable basis, to the corporation in connection with
the establishment and operation of the system.

(c)   the Federal Communications Commission, in its administration of the provisions of the Communications Act of 1934,
as amended, and as supplemented by this chapter, shall--
(1) insure effective competition, including the use of
competitive bidding where appropriate, in the procurement
by the corporation and communications common carriers of
apparatus, equipment, and services required for the establishment and operation of the communications satellite system and satellite terminal stations; and the Commission
shall consult with the Small Business Administration and
solicit its recommendations on measures and procedures which
will insure that small business concerns are given an equitable opportunity to share in the procurement program of the
corporation for property and services, including but not
limited to research, development, construction, maintenance,
and repair.
(2) insure that all present and future authorized carriers
shall have nondiscriminatory use of, and equitable access
to, the communications satellite system and satellite terminal stations under just and reasonable charges, classific-
tions, practices, regulations, and other terms and conditions and regulate the manner in which available facilities
of the system and stations are allocated among such users
thereof;
(3) in any case where the Secretary of State, after obtaining the advice of the Administration as to technical
feasibility, has advised that commercial communication to
a particular foreign point by means of the communications
satellite system and satellite terminal stations should be
established in the national interest, institute forthwith
appropriate proceedings under section 214(d) of this title
to require the establishment of such communication by the
corporation and the appropriate common carrier or carriers;
(4) insure that facilities of the communications satellite system and satellite terminal stations are technically
compatible and interconnected operationally with each other
and with existing communications facilities;
(5) prescribe such accounting regulations and systems
and engage in such ratemaking procedures as will insure
that any economies made possibly by a communications satellite system are appropriately reflected in rates for public
communication services;
(6) approve technical characteristics of the operational
communications satellite system to be employed by the corporation and of the satellite terminal stations; and
(7) grant appropriate authorizations for the construction and operation of each satellite terminal station;
either to the corporation or to one or more authorized
carriers or to the corporation and one or more such carriers jointly, as will best serve the public interest, convenience, and necessity the Commission shall authorize the

construction and operation of such stations by communications common carriers or the corporation, without preference to either;

(8) authorize the corporation to issue any shares of capital stock, except the initial issue of capital stock referred to in section 734(a) of this title, or to borrow any moneys, or to assume any obligation in respect of the securities of any other person, upon a finding that such issuance, borrowing, or assumption is compatible with the public interest, convenience, and necessity and is necessary or appropriate for or consistent with carrying out the purposes and objectives of this chapter by the corporation;

(9) insure that no substantial additions are made by the corporation or carriers with respect to facilities of the system or satellite terminal stations unless such additions are required by the public interest, convenience, and necessity;

(10) require, in accordance with the procedural requirements of section 214 of this title, that additions be made by the corporation or carriers with respect to facilities of the system or satellite terminal stations where such additions would serve the public interest, convenience, and necessity; and

(11) make rules and regulations to carry out the provisions of this chapter.

§ 735. *Powers of corporation; specific activities of corporation; possession of usual powers of stock corporation*

(a) In order to achieve the objectives and to carry out the purposes of this chapter, the corporation is authorized to--

(1) plan, initiate, construct, own, manage, and operate itself or in conjunction with foreign governments or business entities a commercial communications satellite system;

(2) furnish, for hire, channels of communication to United States communications common carriers and to other authorized entities, foreign and domestic; and

(3) own and operate satellite terminal stations when licensed by the Commission under section 721(c)(7) of this title.

(b) Included in the activities authorized to the corporation for accomplishment of the purposes indicated in subsection (a) of this section, are, among others not specifically named--

(1) to conduct or contract for research and development related to its mission;

(2) to acquire the physical facilities, equipment and devices necessary to its operations, including communications satellites and associated equipment and facilities, whether by construction, purchase, or gift;

(3) to purchase satellite launching and related services from the United States Government;

(4) to contract with authorized users, including the United States Government, for the services of the communications satellite system; and

(5) to develop plans for the technical specifications of all elements of the communications satellite system.

(c) To carry out the foregoing purposes, the corporation shall have the usual powers conferred upon a stock corporation by the District of Columbia Business Corporation Act.

U.S. v. Southwestern Cable Co.

In this case the Supreme Court held that the FCC had jurisdiction to regulate CATV systems. Footnotes and various citations have been removed from the Court opinion.

OPINION OF THE COURT

Mr. Justice Harlan delivered the opinion of the Court.

These cases stem from proceedings conducted by the Federal Communications Commission after requests by Midwest Television for relief under §§ 74.1107 and 74.1109 of the rules promulgated by the Commission for the regulation of community antenna television (CATV) systems. Midwest averred that respondents' CATV systems transmitted the signals of Los Angeles broadcasting stations into the San Diego area, and thereby had, inconsistently with the public interest, adversely affected Midwest's San Diego station. Midwest sought an appropriate order limiting the carriage of such signals by respondents' systems. After consideration of the petition and of various responsive pleadings, the Commission restricted the expansion of respondents' service in areas in which they had not operated on February 15, 1966, pending hearings to be conducted on the merits of Midwest's complaints.

On petitions for review, the Court of Appeals for the Ninth Circuit held that the Commission lacks authority under the Communications Act of 1934, 48 Stat 1064, 47 USC § 151, to issue such an order. 378 F2d 118. We granted certiorari to consider this important question of regulatory authority. For reasons that follow, we reverse.

* * *

II.

We must first emphasize that questions as to the validity of the specific rules promulgated by the Commission for the regulation of CATV are not now before the Court. The issues in these cases are only two: whether the Commission has authority under the Communications Act to regulate CATV systems, and, if it has, whether it has, in addition, authority to issue the prohibitory order here in question.

The Commission's authority to regulate broadcasting and other communications is derived from the Communications Act

of 1934, as amended. The Act's provisions are explicitly applicable to "all interstate and foreign communication by wire
or radio . . . ." 47 USC § 152(a). The Commission's responsibilities are no more narrow: it is required to endeavor to
"make available . . . to all the people of the United States
a rapid, efficient, Nation-wide, and world-wide and radio communication service . . . ." 47 USC § 151. The Commission was
expected to serve as the "single Government agency" with "unified jurisdiction" and "regulatory power over all forms of
electrical communication, whether by telephone, telegraph,
cable, or radio." It was for this purpose given "broad authority." As this Court emphasized in an earlier case, the Act's
terms, purposes, and history all indicate that Congress "formulated a unified and comprehensive regulatory system for the
(broadcasting) industry."

* * *

Respondents do not suggest that CATV systems are not within
the term "communication by wire or radio." Indeed, such communications are defined by the Act so as to encompass "the
transmission of . . . signals, pictures, and sounds of all
kinds," whether by radio or cable, "including all instrumentalities, facilities, apparatus, and services (among other
things, the receipt, forwarding, and delivery of communications) incidental to such transmission." 47 USC §§ 153(a),
(b). These very general terms amply suffice to reach respondents' activities.

Nor can we doubt that CATV systems are engaged in interstate
communication, even where, as here, the intercepted signals
emanate from stations located within the same State in which
the CATV system operates. We may take notice that television
broadcasting consists in very large part of programming devised for, and distributed to, national audiences; respondents
thus are ordinarily employed in the simultaneous retransmis-
sion of communications that have very often originated in other
States. The stream of communication is essentially uninter-
rupted and properly indivisible. To categorize respondents'
activities as intrastate would disregard the character of the
television industry, and serve merely to prevent the national
regulation that "is not only appropriate but essential to the
efficient use of radio facilities."

* * *

Nonetheless, respondents urge that the Communications
Act, properly understood, does not permit the regulation of
CATV systems. First, they emphasize that the Commission in
1959 and again in 1966 sought legislation that would have explicitly authorized such regulation, and that its efforts were
unsuccessful. In the circumstances here, however, this cannot
be dispositive. The Commission's requests for legislation
evidently reflected in each instance both its uncertainty as
to the proper width of its authority and its understandable

preference for more detailed policy guidance than the Communications Act now provides. We have recognized that administrative agencies should, in such situations, be encouraged to
seek from Congress clarification of the pertinent statutory
provisions.

* * *

Nor can we obtain significant assistance from the various
expressions of congressional opinion that followed the Commission's requests. In the first place, the views of one Congress
as to the construction of a statute adopted many years before
by another Congress have "very little, if any, significance."

* * *

Further, it is far from clear that Congress believed, as it
considered these requests for legislation, that the Commission
did not already possess regulatory authority over CATV. In
1959, the proposed legislation was preceded by the Commission's
declarations that it "did not intend to regulate CATV," and
that it preferred to recommend the adoption of legislation that
would impose specified requirements upon CATV systems. Congress may well have been more troubled by the Commission's
unwillingness to regulate than by any fears that it was unable
to regulate. In 1966, the Commission informed Congress that
it desired legislation in order to "confirm (its) jurisdiction
and to establish such basic national policy as (Congress) deems
appropriate." HR Rep No. 1635, 89th Cong, 2d Sess, 16. In response, the House Committee on Interstate and Foreign Commerce
said merely that it did not "either agree or disagree" with the
jurisdictional conclusions of the Second Report, and that "the
question of whether or not . . . the Commission has authority
under present law to regulate CATV systems is for the courts
to decide . . . ." Id., at 9. In these circumstances, we cannot derive from the Commission's requests for legislation anything of significant bearing on the construction question now
before us.

Second, respondents urge that § 152(a) does not independently confer regulatory authority upon the Commission, but
instead merely prescribes the forms of communications to which
the Act's other provisions may separately be made applicable.
Respondents emphasize that the Commission does not contend
either that CATV systems are common carriers, and thus within
Title II of the Act, or that they are broadcasters, and thus
within Title III. They conclude that CATV, with certain of the
characteristics both of broadcasting and of common carriers,
but with all of the characteristics of neither, eludes altogether the Act's grasp.

We cannot construe the Act so restrictively. Nothing in the
language of § 152(a), in the surrounding language, or in the
Act's history or purposes limits the Commission's authority to
those activities and forms of communication that are specifically described by the Act's other provisions. The section it-

self states merely that the "provisions of (the Act) shall
apply to all interstate and foreign communication by wire or
radio . . . ." Similarly, the legislative history indicates
that the Commission was given "regulatory power over all forms
of electrical communication . . . ." S Rep No. 781, 73d Cong,
2d Sess, 1. Certainly Congress could not in 1934 have foreseen
the development of community antenna television systems, but it
seems to us that it was precisely because Congress wished "to
maintain, through appropriate administrative control, a grip on
the dynamic aspects of radio transmission," that it conferred
upon the Commission a "unified jurisdiction" and "broad authority." Thus, "(u)nderlying the whole (Communications Act) is
recognition of the rapidly fluctuating factors characteristic
of the evolution of broadcasting and of the corresponding requirement that the administrative process possess sufficient
flexibility to adjust itself to these factors."
                              * * *
   Congress in 1934 acted in a field that was demonstrably
"both new and dynamic," and it therefore gave the Commission
"a comprehensive mandate," with "not niggardly but expansive
powers."
                              * * *
   We have found no reason to believe that § 152 does not, as
its terms suggest, confer regulatory authority over "all interstate . . . communication by wire or radio."
   Moreover, the Commission has reasonably concluded that regulatory authority over CATV is imperative if it is to perform
with appropriate effectiveness certain of its other responsibilities. Congress has imposed upon the Commission the "obligation of providing a widely dispersed radio and television,
service," with a "fair, efficient, and equitable distribution"
of service among the "several States and communities." 47 USC
§ 307(b). The Commission has, for this and other purposes,
been granted authority to allocate broadcasting zones or areas,
and to provide regulations "as it may deem necessary" to prevent interference among the various stations. 47 USC §§ 303(f),
(h). The Commission has concluded, and Congress has agreed,
that these obligations require for their satisfaction the creation of a system of local broadcasting stations, such that "all
communities of appreciable size (will) have at least one television station as an outlet for local self-expression." In
turn, the Commission has held that an appropriate system of
local broadcasting may be created only if two subsidiary goals
are realized. First, significantly wider use must be made of
the available ultra-high-frequency channels. Second, communities must be encouraged "to launch sound and adequate programs
to utilize the television channels now reserved for educational
purposes." These subsidiary goals have received the endorse-
ment of Congress.

The Commission has reasonably found that the achievement of
each of these purposes is "placed in jeopardy by the unregu-
lated explosive growth of CATV." HR Rep No. 1635, 89th Cong,
2d Sess, 7. Although CATV may in some circumstances make possible "the realization of some of the (Commission's) most
important goals,"

* * *

its importation of distant signals into the service areas
of local stations may also "destroy or seriously degrade the
service offered by a television broadcaster," id., at 700, and
thus ultimately deprive the public of the various benefits of
a system of local broadcasting stations. In particular, the
Commission feared that CATV might, by dividing the available
audiences and revenues, significantly magnify the characteristically serious financial difficulties of UHF and educational television broadcasters. The Commission acknowledged that
it could not predict with certainty the consequences of unregu-
lated CATV, but reasoned that its statutory responsibilities
demand that it "plan in advance of foreseeable events, instead
of waiting to react to them."

We are aware that these consequences have been variously
estimated, but must conclude that there is substantial evidence that the Commission cannot "discharge its overall responsibilities without authority over this important aspect of
television service." Staff of Senate Comm on Interstate and
Foreign Commerce, 85th Cong, 2d Sess, The Television Inquiry:
The Problem of Television Service for Smaller Communities 19
(Comm Print 1959).

The Commission has been charged with broad responsibilities
for the orderly development of an appropriate system of local
television broadcasting. The significance of its efforts can
scarcely be exaggerated, for broadcasting is demonstrably a
principal source of information and entertainment for a great
part of the Nation's population. The Commission has reasonably found that the successful performance of these duties demands prompt and efficacious regulation of community antenna
television systems. We have elsewhere held that we may not,
"in the absence of compelling evidence that such was Congress'
intention . . . prohibit administrative action imperative for
the achievement of any agency's ultimate purposes."

* * *

There is no such evidence here, and we therefore hold that
the Commission's authority over "all interstate . . . communication by wire or radio" permits the regulation of CATV systems.

There is no need here to determine in detail the limits of
the Commission's authority to regulate CATV. It is enough to
emphasize that the authority which we recognize today under
§ 152(a) is restricted to that reasonably ancillary to the effective performance of the Commission's various responsibili-

ties for the regulation of television broadcasting. The Commission may, for these purposes, issue "such rules and regulations and prescribe such restrictions and conditions, not
inconsistent with law," as "public convenience, interest, or
necessity requires." 47 USC § 303(r). We express no views as
to the commission's authority, if any, to regulate CATV under
any other circumstances or for any other purposes.

                            * * *

FCC Regulations

What follows are selected portions of the regulations formulated by the Federal Communications Commission in response to
the All-Channel Receiver Act and the *Southwestern Cable* decision. These are found in 47 C.F.R. at the appropriate section
indicated in the excerpt.

§ *15.65 All-channel television broadcast reception: General*
          requirement.
Except as provided in § 15.66, all television broadcast receivers manufactured after April 30, 1964, and shipped in
interstate commerce or imported from any foreign country into
the United States, for sale or resale to the public, shall be
capable of adequately receiving all channels allocated by the
Commission to the television broadcast service. A television
broadcast receiver is capable of adequately receiving all
channels if it meets the requirements of §§ 15.67 and 15.68
in effect on the day of its manufacture.

§ *15.67 All-channel television broadcast reception: Receivers*
          manufactured after April 30, 1964.
Television broadcast receivers manufactured after April 30,
1964, shall comply with the following specifications:
    (a) *Noise figure.* The noise figure for any television chan-
nel between 14 and 83 inclusive, shall not be larger than 18
db.
    (b) *Peak picture sensitivity.* The peak picture sensitivity
of any television broadcast receiver, average for all channels
between 14 and 83 inclusive, shall not be more than 8 db larger
than the peak picture sensitivity of that receiver averaged for
all television channels between 2 and 13 inclusive. (35 F.R.
2666, Feb. 6, 1970)

§ *15.68 All-channel television broadcast reception: Receivers*
          manufactured on or after July 1, 1971.
    (a) *Effective date.* The requirements of this section, in
addition to the requirements of § 15.67, shall apply to 10 percent of the television receiver models produced by any domestic
manufacturer, or exported to the United States by any foreign
manufacturer, on or after July 1, 1971; 40 percent of the models produced (or exported to the United States) by any manufacturer on or after July 1, 1972; 70 percent of the models

produced (or exported to the United States) by any manufacturer on or after July 1, 1973; and to all receivers manufactured (or exported to the United States) on or after July 1,
1974. They shall, in addition, apply to any receiver model
manufactured (or exported to the United States) on or after
January 1, 1972, and not manufactured prior to that date.

*Note: The term "model" refers to all of a type of television broad-
cast receiver made (or exported to the United States) by a single manufacturer which combines the same basic chassis with the same size picture
tube. To determine the number of models subject to the requirements on
the interim compliance dates, multiply the total number of models by the
appropriate percentage and reduce the result to the next lowest whole
number.*

(b) *Receiver requirements.* On a given receiver (after any
initial adjustment of a detent mechanism required to receive
UHF channels), use of the UHF and VHF tuning systems shall provide approximately the same degree of tuning accuracy with approximately the same expenditure of time and effort.

(1) *Basic tuning mechanism.* If any television receiver
is equipped to provide for repeated access to VHF television
channels at discrete tuning positions, that receiver shall
be equipped to provide for repeated access to a minimum of
six UHF television channels at discrete tuning positions.
Unless a discrete tuning position is provided for each
channel allocated to UHF television, each position shall be
readily adjustable to a particular UHF channel by the user
without the use of tools. If 12 or fewer discrete tuning
positions are provided, each position shall be adjustable
to receive any channel allocated to UHF television.

*Note: The combination of detented rotary switch and pushbutton con-
trols is acceptable, provided UHF channels, after their initial selection, can be accurately tuned with an expenditure of time and effort
approximately the same as that used in accurately tuning VHF channels.
A UHF tuning system comprising five pushbuttons and a separate manual
tuning knob is considered to provide repeated access to six channels
at discrete tuning positions. A one-knob (VHF-UHF) tuning system providing repeated access to 11 or more discrete tuning positions is also
acceptable, provided each of the tuning positions is readily adjustable,
without use of tools, to receive any UHF channel.*

(2) *Tuning aids.* If equipment and controls which tend
to simplify, expedite or perfect the reception of television signals (e.g., AFC, visual aids, remote control, or
signal seeking capability, referred to generally as tuning
aids) are incorporated into the design of a television
broadcast receiver, tuning aids of the same type and of
comparable capability and quality shall be provided for
tuning both the VHF television channels and the UHF television channels.

(3) *Tuning controls and channel read-out.* UHF tuning
controls and channel read-out on a given receiver shall be

comparable in size, location, accessibility and legibility
to VHF tuning controls and readout on that receiver. If
any television receiver utilizes continuous UHF tuning for
any function (e.g., as the basic tuning mode, for preset-
ting a detent mechanism for repeated access at discrete
tuning positions, or for tuning a channel which cannot be
assigned a discrete tuning position), that receiver shall
be equipped to display the approximate UHF television channel the tuner has been positioned to receive. If any television receiver is equipped to provide repeated access to
UHF television channels at discrete tuning positions, the
manufacturer shall provide for the display of the precise
UHF channel selected or shall provide to the user a means
of identifying the precise channel selected without the
use of tools: *Provided, however,* That the 70 UHF channel
numbers may be displayed in groups of three or less at each
of 24 settings, if

(i) The tuning mechanism uses a single control to select the VHF and UHF channels;

(ii) Any one of the three channels simultaneously displayed can be precisely tuned to the correct frequency;
and

(iii) The reset accuracy (with AFC, if provided) is
sufficient to eliminate the need for routine fine tuning.

(c) *Progress reports.* Television receiver manufacturers
shall file periodic reports detailing their progress in meeting the requirements of this section. The reports shall be
filed regularly, on June 1 and December 1 of each year, beginning on December 1, 1970, and shall be directed to the Office
of Chief Engineer, Federal Communications Commission, Washington, D.C. 20554. Any manufacturer who expects to encounter
difficulty in meeting the schedule for compliance shall, in
addition, at the earliest possible date, file a special report
detailing the difficulties encountered and the steps being
taken to overcome them.

(d) *Use of a 70-position nonmemory UHF detent tuning system.*

(1) Numerical readout shall be provided for each of the
70 UHF channels, or, if all of the VHF and UHF channel numbers displayed are at all times visible on the face of the
receiver, numerical readout shall be provided for at least
every other UHF channel, with marks to indicate those channels not displayed numerically.

(2) Until July 1, 1975, a 70-position, nonmemory UHF
detent tuning system may be used to meet the requirements
of this section provided the channel selection mechanism
shall be capable of positioning the tuner to receive each
UHF channel at its designated detent position, with maximum deviation from correct frequency on any detent setting
not exceeding ±3 MHz, when approached from either direction
of rotation.

(3) On or after July 1, 1975, a 70-position nonmemory
UHF detent tuning system may be used to meet the requirements of this section provided the channel selection mechanism shall be capable of positioning the tuner to receive
each UHF channel at its designated detent position, with
maximum deviation from correct frequency on any detent setting not exceeding ±2 MHz, when approached from either direction of rotation.

(4) On or after July 1, 1976, a 70-position nonmemory
UHF detent tuning system may be used to meet the requirements of this section, providing either of the following
two conditions is met:

(i) *For any television receiver (monochrome or color).*
The need for routine fine tuning of UHF channels is
eliminated.

*Note: This requirement will be considered met in each of the follow-
ing circumstances:*

*The receiver is provided with AFC and a channel selection mechanism
that is capable of positioning the tuner to receive each UHF channel
at its designated detent position with a maximum deviation from correct frequency on any detent setting not exceeding ±1 MHz, when approached from either direction of rotation.*

*The receiver is provided with AFC and a channel selection mechanism
that is capable of positioning the tuner to receive each UHF channel
at its designated detent position within the pull in range of the AFC,
when approached from either direction of rotation.*

*The receiver is provided with any other tuning system that produces
and maintains detented tuning accuracy of the same order as the above
specified systems.*

(ii) *For monochrome receivers only.* The UHF channel
selection mechanism is capable of positioning the tuner
to receive each UHF channel at its designated detent
position, with maximum deviation from correct frequency
on any detent setting not exceeding ±1 MHz, when approached from either direction of rotation.

§ *76.251 Minimum channel capacity; access channels.*

(a) No cable television system operating in a community located in whole or in part within a major television market, as
defined in § 76.5, shall carry the signal of any television
broadcast station unless the system also complies with the following requirements concerning the availability and administration of access channels:

(1) *Minimum channel capacity.* Each such system shall
have at least 120 MHz of bandwidth (the equivalent of 20
television broadcast channels) available for immediate or
potential use for the totality of cable services to be offered;

(2) *Equivalent amount of bandwidth.* For each Class I
cable channel that is utilized, such system shall be capable of providing an additional channel, 6 MHz in width,

suitable for transmission of Class II or Class III signals
(see § 76.5 for cable channel definitions);

(3) *Two-way communications.* Each such system shall
maintain a plant having technical capacity for nonvoice return communications;

(4) *Public access channel.* Each such system shall maintain at least one specially designated, noncommercial public
access channel available on a first-come, nondiscriminatory
basis. The system shall maintain and have available for
public use at least the minimal equipment and facilities
necessary for the production of programing for such a channel. See also § 76.201;

(5) *Education access channel.* Each such system shall
maintain at least one specially designated channel for use
by local educational authorities;

(6) *Local government access channel.* Each such system
shall maintain at least one specially designated channel
for local government uses;

(7) *Leased access channels.* Having satisfied the origination cablecasting requirements of § 76.201, and the requirements of subparagraphs (4), (5), and (6) of this paragraph for specially designated access channels, such system
shall offer other portions of its nonbroadcast bandwidth,
including unused portions of the specially designated channels, for leased access services. However, these leased
channel operations shall be undertaken with the express
understanding that they are subject to displacement if there
is a demand to use the channels for their specially designated purposes. On at least one of the leased channels,
priority shall be given part-time users;

(8) *Expansion of access channel capacity.* Whenever all
of the channels described in subparagraphs (4) through (7)
of this paragraph are in use during 80 percent of the week-
days (Monday-Friday) for 80 percent of the time during any
consecutive 3-hour period for 6 consecutive weeks, such system shall have 6 months in which to make a new channel
available for any or all of the above-described purposes;

(9) *Program content control.* Each such system shall
exercise no control over program content on any of the channels described in subparagraphs (4) through (7) of this
paragraph; however, this limitation shall not prevent it
from taking appropriate steps to insure compliance with the
operating rules described in subparagraph (11) of this paragraph;

(10) *Assessment of costs.*

(i) From the commencement of cable television service
in the community of such system until five (5) years
after completion of the system's basic trunk line, the
channels described in subparagraphs (5) and (6) of this
paragraph shall be made available without charge.

(ii) One of the public access channels described in subparagraph (4) of this paragraph shall always be made available without charge, except that production costs may be assessed for live studio presentations exceeding 5 minutes. Such production costs and any fees for use of other public access channels shall be consistent with the goal of affording the public a low-cost means of television access;

(11) *Operating rules.*

(i) For the public access channel(s), such system shall establish rules requiring first-come nondiscriminatory access; prohibiting the presentation of: Any advertising material designed to promote the sale of commercial products or services (including advertising by or on behalf of candidates for public office); lottery information; and obscene or indecent matter (modeled after the prohibitions in §§ 76.213 and 76.215, respectively); and permitting public inspection of a complete record of the names and addresses of all persons or groups requesting access time. Such a record shall be retained for a period of 2 years.

(ii) For the educational access channel(s), such system shall establish rules prohibiting the presentation of: Any advertising material designed to promote the sale of commercial products or services (including advertising by or on behalf of candidates for public office); lottery information; and obscene or indecent matter (modeled after the prohibitions in §§ 76.213 and 76.215, respectively) and permitting public inspection of a complete record of the names and addresses of all persons or groups requesting access time. Such a record shall be retained for a period of 2 years.

(iii) For the leased channel(s), such system shall establish rules requiring first-come, nondiscriminatory access; prohibiting the presentation of lottery information and obscene or indecent matter (modeled after the prohibitions in §§ 76.213 and 76.215, respectively); requiring sponsorship identification (see § 76.221); specifying an appropriate rate schedule and permitting public inspection of a complete record of the names and addresses of all persons or groups requesting time. Such a record shall be retained for a period of 2 years.

(iv) The operating rules governing public access, educational, and leased channels shall be filed with the Commission within 90 days after a system first activates any such channels, and shall be available for public inspection at the system's offices. Except on specific authorization, or with respect to the operation of the local government access channel, no local entity shall prescribe any other rules concerning the number or man-

ner of operation of access channels; however, franchise
specifications concerning the number of such channels
for systems in operation prior to March 31, 1972, shall
continue in effect.

(b) No cable television system operating in a community located wholly outside of all major television markets shall be
required by a local entity to exceed the provisions concerning
the availability and administration of access channels contained in paragraph (a) of this section. If a system provides
any access programing, it shall comply with paragraph (a) (9),
(10), and (11) of this section.

(c) The provisions of this section shall apply to all cable
television systems that commence operations on or after March
31, 1972, in a community located in whole or in part within a
major television market. Systems that commenced operations
prior to March 31, 1972, shall comply on or before March 31,
1977: *Provided, however,* That, if such systems begin to provide
any of the access services described above at an earlier date,
they shall comply with paragraph (a)(9), (10), and (11) of this
section at that time: *And provided, further,* That if such systems receive certificates of compliance to add television signals to their operations at an earlier date, pursuant to
§ 76.61(b) or (c), or § 76.63(a) (as it relates to § 76.61(b)
or (c)), for each such signal added, such systems shall provide
one (1) access channel in the following order of priority--(1)
public access, (2) education access, (3) local government
access, and (4) leased access--and shall comply with the appropriate requirements of paragraphs (a) (4)-(7) and (a) (9)-(11)
of this section with respect thereto.

§ *76.51 Major television markets.*
For purposes of the cable television rules, the following is
a list of the major television markets and their designated
communities:

(a) First 50 major television markets:

(1) New York, N.Y.--Linden-Paterson, N.J.
(2) Los Angeles-San Bernardino-Corona-Fontant, Calif.
(3) Chicago, Ill.
(4) Philadelphia, Pa.-Burlington, N.J.
(5) Detroit, Mich.
(6) Boston-Cambridge-Worcester, Mass.
(7) San Francisco-Oakland-San Jose, Calif.
(8) Cleveland-Lorain-Akron, Ohio.
(9) Washington, D.C.
(10) Pittsburgh, Pa.
(11) St. Louis, Mo.
(12) Dallas-Fort Worth, Tex.
(13) Minneapolis-St. Paul, Minn.
(14) Baltimore, Md.
(15) Houston, Tex.
(16) Indianapolis-Bloomington, Ind.
(17) Cincinnati, Ohio-Newport, Ky.

(18) Atlanta, Ga.
(19) Hartford-New Haven-New Britain-Waterbury, Conn.
(20) Seattle-Tacoma, Wash.
(21) Miami, Fla.
(22) Kansas City, Mo.
(23) Milwaukee, Wis.
(24) Buffalo, N.Y.
(25) Sacramento-Stockton-Modesto, Calif.
(26) Memphis, Tenn.
(27) Columbus, Ohio.
(28) Tampa-St. Petersburg, Fla.
(29) Portland, Oreg.
(30) Nashville, Tenn.
(31) New Orleans, La.
(32) Denver, Colo.
(33) Providence, R.I.-New Bedford, Mass.
(34) Albany-Schenectady-Troy, N.Y.
(35) Syracuse, N.Y.
(36) Charleston-Huntington, W. Va.
(37) Kalamazoo-Grand Rapids-Muskegon-Battle Creek, Mich.
(38) Louisville, Ky.
(39) Oklahoma City, Okla.
(40) Birmingham, Ala.
(41) Dayton-Kettering, Ohio.
(42) Charlotte, N.C.
(43) Phoenix-Mesa, Ariz.
(44) Norfolk-Newport News-Portsmouth-Hampton, Va.
(45) San Antonio, Tex.
(46) Greenville-Spartanburg-Anderson, S.C.-Asheville, N.C.
(47) Greensboro-High Point-Winston Salem, N.C.
(48) Salt Lake City, Utah.
(49) Wilkes Barre-Scranton, Pa.
(50) Little Rock, Ark.

(b) Second 50 major television markets:

(51) San Diego, Calif.
(52) Toledo, Ohio.
(53) Omaha, Nebr.
(54) Tulsa, Okla.
(55) Orlando-Daytona Beach, Fla.
(56) Rochester, N.Y.
(57) Harrisburg-Lancaster-York, Pa.
(58) Texarkana, Tex.-Shreveport, La.
(59) Mobile, Ala.-Pensacola, Fla.
(60) Davenport, Iowa-Rock Island-Moline, Ill.
(61) Flint-Bay City-Saginaw, Mich.
(62) Green Bay, Wis.
(63) Richmond-Petersburg, Va.
(64) Springfield-Decatur-Champaign-Jacksonville, Ill.
(65) Cedar Rapids-Waterloo, Iowa.
(66) Des Moines-Ames, Iowa.
(67) Wichita-Hutchinson, Kans.
(68) Jacksonville, Fla.
(69) Cape Girardeau, Mo.-Paducah, Ky.-Harrisburg, Ill.
(70) Roanoke-Lynchburg, Va.
(71) Knoxville, Tenn.

(72) Fresno, Calif.
(73) Raleigh-Durham, N.C.
(74) Johnstown-Altoona, Pa.
(75) Portland-Poland Spring, Maine.
(76) Spokane, Wash.
(77) Jackson, Miss.
(78) Chattanooga, Tenn.
(79) Youngstown, Ohio.
(80) South Bend-Elkhart, Ind.
(81) Albuquerque, N. Mex.
(82) Fort Wayne-Roanoke, Ind.
(83) Peoria, Ill.
(84) Greenville-Washington-New Bern, N.C.
(85) Sioux Falls-Mitchell, S. Dak.
(86) Evansville, Ind.
(87) Baton Rouge, La.
(88) Beaumont-Port Arthur, Tex.
(89) Duluth, Minn.-Superior, Minn.(sic)
(90) Wheeling, W. Va.-Steubenville, Ohio.
(91) Lincoln-Hastings-Kearney, Nebr.
(92) Lansing-Onondaga, Mich.
(93) Madison, Wis.
(94) Columbus, Ga.
(95) Amarillo, Tex.
(96) Huntsville-Decatur, Ala.
(97) Rockford-Freeport, Ill.
(98) Fargo-Valley City, N.D.
(99) Monroe, La.-El Dorado, Ark.
(100) Columbia, S.C.

§ 76.57 *Provisions for systems operating in communities located outside of all major and smaller television
markets.*
A cable television system operating in a community located
wholly outside all major and smaller television markets, as
defined in § 76.5, shall carry television broadcast signals
in accordance with the following provisions:
  (a) Any such cable television system may carry or, on request of the relevant station licensee or permittee, shall
carry the signals of:
    (1) Television broadcast stations within whose Grade B
  contours the community of the system is located, in whole
  or in part;
    (2) Television translator stations with 100 watts or
  higher power serving the community of the system and, as
  to cable systems that commence operations or expand channel capacity after March 30, 1972, noncommercial educational
  translator stations with 5 watts or higher power serving the
  community of the system. In addition, any cable system may
  elect to carry the signal of any noncommercial educational
  translator station;
    (3) Noncommercial educational television broadcast sta-
  tions within whose specified zone the community of the system is located, in whole or in part;

(4) Commercial television broadcast stations that are
significantly viewed in the community of the system. See
§ 76.54.
(b) In addition to the television broadcast signals carried
pursuant to paragraph (a) of this section, any such cable television system may carry any additional television signals.

§ 76.59 *Provisions for smaller television markets.*
A cable television system operating in a community located in
whole or in part within a smaller television market, as defined
in § 76.5, shall carry television broadcast signals only in accordance with the following provisions:
(a) Any such cable television system may carry or, on request of the relevant station licensee or permittee, shall
carry the signals of:
   (1) Television broadcast stations within whose specified
   zone the community of the system is located, in whole or in
   part;
   (2) Noncommercial educational television broadcast stations within whose Grade B contours the community of the
   system is located, in whole or in part;
   (3) Commercial television broadcast stations licensed to
   communities in other smaller television markets, within
   whose Grade B contours the community of the system is located, in whole or in part;
   (4) Television broadcast stations licensed to other communities which are generally considered to be part of the
   same smaller television market (Example: Burlington, Vt.-
   Plattsburgh, N.Y., television market);
   (5) Television translator stations with 100 watts or
   higher power serving the community of the system and, as
   to cable systems that commence operations or expand channel
   capacity after March 30, 1972, noncommercial educational
   translator stations with 5 watts or higher power serving
   the community of the system. In addition, any cable system
   may elect to carry the signal of any noncommercial educational translator station;
   (6) Commercial television broadcast stations that are
   significantly viewed in the community of the system. See
   § 76.54.
(b) Any such cable television system may carry sufficient
additional signals so that, including the signals required to
be carried pursuant to paragraph (a) of this section, it can
provide the signals of a full network station of each of the
major national television networks, and of one independent
television station: *Provided, however,* That, in determining
how many additional signals may be carried, any authorized but
not operating television broadcast station that, if operational,
would be required to be carried pursuant to paragraph (a)(1) of
this section, shall be considered to be operational for a peri-

od terminating 18 months after grant of its initial construction permit. The following priorities are applicable to the additional television signals that may be carried:

(1) *Full network stations.* A cable television system may carry the nearest full network stations or the nearest in-state full network stations;

Note: The Commission may waive the requirements of this subparagraph for good cause shown in a petition filed pursuant to § 76.7.

(2) *Independent station.* A cable television system may carry any independent television station: *Provided, however,* That if a signal of a station in the first 25 major television markets (see § 76.51(a)) is carried pursuant to this subparagraph, such signal shall be taken from one of the two closest such markets, where such signal is available.

Note: It is not contemplated that waiver of the provisions of this subparagraph will be granted.

(c) In addition to the noncommercial educational television broadcast signals carried pursuant to paragraph (a) of this section, any such cable television system may carry the signals of any noncommercial educational stations that are operated by an agency of the State within which the system is located. Such system may also carry any other noncommercial educational signals, in the absence of objection filed pursuant to § 76.7 by any local noncommercial educational station or State or local educational television authority.

(d) In addition to the television broadcast signals carried pursuant to paragraphs (a) through (c) of this section, any such cable television system may carry:

(1) Any television stations broadcasting predominantly in a non-English language; and

(2) Any television station broadcasting a network program that will not be carried by a station normally carried on the system. Carriage of such additional stations shall be only for the duration of the network programs not otherwise available, and shall not require prior Commission notification or approval in the certificating process.

(e) Where the community of a cable television system is wholly or partially within both one of the first 50 major television markets and a smaller television market, the carriage provisions for the first 50 major markets shall apply. Where the community of a system is wholly or partially within both one of the second 50 major television markets and a smaller television market, the carriage provisions for the second 50 major markets shall apply.

(37 F.R. 3278, Feb. 12, 1972, as amended at 37 F.R. 13867, July 14, 1972)

§ *76.61   Provisions for first 50 major television markets.*
A cable television system operating in a community located in
whole or in part within one of the first 50 major television
markets listed in § 76.51(a) shall carry television broadcast
signals only in accordance with the following provisions:

(a) Any such cable television system may carry, or on request of the relevant station licensee or permittee, shall
carry the signals of:

(1) Television broadcast stations within whose specified
zone the community of the system is located, in whole or in
part: *Provided, however,* That where a cable television system is located in the designated community of a major television market, it shall not carry the signal of a television
station licensed to a designated community in another major
television market, unless the designated community in which
the cable system is located is wholly within the specified
zone (see § 76.5(f) of the station, except as otherwise provided in this section;

(2) Noncommercial educational television broadcast stations within whose Grade B contours the community of the
system is located, in whole or in part;

(3) Television translator stations with 100 watts or
higher power serving the community of the system and, as to
cable systems that commence operations or expand channel
capacity after March 30, 1972, noncommercial educational
translator stations with 5 watts or higher power serving
the community of the system. In addition, any cable system
may elect to carry the signal of any noncommercial educational translator station;

(4) Television broadcast stations licensed to other designated communities of the same major television market (Example: Cincinnati, Ohio-Newport, Ky., television market);

(5) Commercial television broadcast stations that are
significantly viewed in the community of the system. See
§ 76.54.

(b) Any such cable television system may carry sufficient
additional signals so that, including the signals required to
be carried pursuant to paragraph (a) of this section, it can
provide the signals of a full network station of each of the
major national television networks, and of three independent
television stations; *Provided, however,* That in determining
how many additional signals may be carried, any authorized but
not operating television broadcast station that, if operational,
would be required to be carried pursuant to paragraph (a)(1) of
this section, shall be considered to be operational for a period terminating 18 months after grant of its initial construction
permit. The following priorities are applicable to the additional television signals that may be carried:

(1) *Full network stations.* A cable television system
may carry the nearest full network stations, or the nearest
in-State full network stations;

*Note: The Commission may waive the requirements of this subparagraph
for good cause shown in a petition filed pursuant to § 76.7.*

(2) *Independent stations.*

(i) For the first and second additional signals, if
any, a cable television system may carry the signals of
any independent television station: *Provided, however,*
That if signals of stations in the first 25 major television markets (see § 76.51(a)) are carried pursuant to
this subparagraph, such signals shall be taken from one
or both of the two closest such markets, where such signals are available. If a third additional signal may be
carried, a system shall carry the signal of any independent UHF television station located within 200 air miles
of the reference point for the community of the system
(see § 76.53), or, if there is no such station, either
the signal of any independent VHF television station
located within 200 air miles of the reference point for
the community of the system, or the signal of any independent UHF television station.

*Note: It is not contemplated that waiver of the provisions of this
subparagraph will be granted.*

(ii) Whenever, pursuant to Subpart F of this part, a
cable television system is required to delete a television program on a signal carried pursuant to subdivi-
sion (i) of this subparagraph or paragraph (c) of this
section, or a program on such a signal is primarily of
local interest to the distant community (e.g., a local
news or public affairs program), such system may, consistent with the program exclusivity rules of Subpart
F of this part, substitute a program from any other
television broadcast station. A program substituted may
be carried to its completion, and the cable system need
not return to its regularly carried signal until it can
do so without interrupting a program already in progress.

(c) After the service standards specified in paragraph (b)
of this section have been satisfied, a cable television system
may carry two additional independent television broadcast signals, chosen in accordance with the priorities specified in
paragraph (b)(2) of this section: *Provided, however,* That the
number of additional signals permitted under this paragraph
shall be reduced by the number of signals added to the system
pursuant to paragraph (b) of this section.

(d) In addition to the noncommercial educational television
broadcast signals carried pursuant to paragraph (a) of this
section, any such cable television system may carry the signals
of any noncommercial educational stations that are operated by
an agency of the State within which the system is located.
Such system may also carry any other noncommercial educational
signals, in the absence of objection filed pursuant to § 76.7

by any local noncommercial educational system or State or local
educational television authority.

(e) In addition to the television broadcast signals carried
pursuant to paragraphs (a) through (d) of this section, any
such cable television system may carry:

(1) Any television stations broadcasting predominantly
in a non-English language; and

(2) Any television station broadcasting a network program that will not be carried by a station normally carried on the system. Carriage of such additional stations
shall be only for the duration of the network programs not
otherwise available, and shall not require prior Commission
notification or approval in the certificating process.

(f) Where the community of a cable television system is
wholly or partially within both one of the first 50 major television markets and another television market, the provisions
of this section shall apply.

§ 76.63 *Provisions for second 50 major television markets.*

(a) A cable television system operating in a community located in whole or in part within one of the second 50 major
television markets listed in § 76.51(b) shall carry television
broadcast signals only in accordance with the provisions of
§ 76.61, except that in paragraph (b) of § 76.61, the number
of additional independent television signals that may be carried is two (2).

(b) Where the community of a cable television system is
wholly or partially within both one of the second 50 major
television markets and one of the first 50 major television
markets, the carriage provisions for the first 50 major markets shall apply. Where the community of a system is wholly
or partially within both one of the second 50 major television
markets and a smaller television market, the provisions of
this section shall apply.

FCC Decisions

The following are two of the Federal Communications Commission's decisions which have been discussed in this book. The
excerpts do not include the entire decision and rationale
but only the central issues and the commission's treatment of
them.

BEFORE THE
FEDERAL COMMUNICATIONS COMMISSION
WASHINGTON, D.C. 20554

In the Matter of
Use of the Carterfone Device in Message   *Docket No. 16942*
  Toll Telephone Service

In the Matter of
Thomas F. Carter and Carter Electronics
Corp., Dallas, Tex. (complainants), v.   *Docket No. 17073*
American Telephone and Telegraph Co.,
Associated Bell System Companies,
Southwestern Bell Telephone Co., and
General Telephone Co. of the South-
west (defendants)

DECISION

(Adopted June 26, 1968)

BY COMMISSIONER JOHNSON FOR THE COMMISSION: COMMISSIONER
LOEVINGER DID NOT PARTICIPATE IN THE DECISION OF THIS CASE.

This proceeding involves the application of American Telephone and Telegraph Co. tariffs to the use by telephone subscribers of the Carterfone.

The Carterfone is designed to be connected to a two-way
radio at the base station serving a mobile radio system. When
callers on the radio and on the telephone are both in contact
with the base station operator, the handset of the operator's
telephone is placed on a cradle in the Carterfone device. A
voice control circuit in the Carterfone automatically switches
on the radio transmitter when the telephone caller is speaking;
when he stops speaking, the radio returns to a receiving con-
dition. A separate speaker is attached to the Carterfone to
allow the base station operator to monitor the conversation,
adjust the voice volume, and hang up his telephone when the
conversation has ended.

The Carterfone device, invented by Thomas F. Carter, has
been produced and marketed by the Carter Electronics Corp.,
of which Mr. Carter is president, since 1959. From 1959
through 1966 approximately 4,500 Carterfones were produced and
3,500 sold to dealers and distributors throughout the United
States and in foreign countries.

The defendant telephone companies, acting in accordance
with their interpretation of tariff FCC No. 132, filed April
16, 1957, by American Telephone and Telegraph Co., advised
their subscribers that the Carterfone, when used in conjunction
with the subscriber's telephone, is a prohibited interconnecting device, the use of which would subject the user to the
penalties provided in the tariff. The tariff provides that:

No equipment, apparatus, circuit or device not furnished by the telephone company shall be attached to or connected with the facilities
furnished by the telephone company, whether physically, by induction or
otherwise. * * * (A fuller text is provided in appendix A.)

* * *

On October 20, 1966, the Commission on its own motion or-
dered that a public hearing be held to resolve "the question
of the justness, reasonableness, validity, and effect of the
tariff regulations and practices complained of," assigning

docket No. 16942. The following five specific issues were
designated for hearing:

1. The nature and extent of the public need and demand for the use of
the Carterfone device in connection with interstate or foreign message
toll telephone service;

2. The effect of the use of the Carterfone device upon the operation
of the telephone system used to provide interstate and foreign telephone
message toll telephone services to the public or upon the employees and
facilities of the telephone companies providing such services or upon
the public in its use of such telephone system;

3. Whether the provisions of tariff FCC No. 132 filed by American
Telephone and Telegraph Co. may properly be construed to prohibit any
telephone user from attaching the Carterfone device to the facilities
of the telephone companies for use in connection with interstate and
foreign message toll telephone services;

4. If the aforesaid tariff provisions may properly be construed to
prohibit telephone users from attaching the Carterfone device to the
facilities of the telephone companies for use in connection with interstate or foreign message toll telephone services;

(a) Whether such regulations are, or will be, unjust and unreasonable and, therefore, unlawful within the meaning of section 201(b) of
the Communications Act of 1934, as amended, or are, or will be unduly
discriminatory or preferential in violation of section 202(a) of said
Act;[1]

(b) Whether, in the light of facts developed in connection with
the foregoing issues, the Commission, in accordance with the provisions of section 205 of the Act, should prescribe tariff regulations
which will permit the use of the Carterfone device in connection with
interstate and foreign toll telephone service and, if so, the kind of
tariff regulations which should be prescribed;

5. If the aforesaid tariff regulations of the telephone companies may
not properly be construed to prohibit telephone users from attaching the
Carterfone device to the facilities of the telephone companies for use
in connection with interstate or foreign message toll telephone services,
what action, if any, should be taken by the Commission with respect
thereto.

* * *

By order released March 8, 1967, the complaint was consoli-
dated for hearing with docket No. 16942, and the following
issues were added:

1. Whether, with respect to the period from February 6, 1957, to December 21, 1966, the regulations and practices of tariff FCC No. 132 of
the American Telephone and Telegraph Co. were properly construed and
applied to prohibit any telephone user from attaching the Carterfone
device to the facilities of the telephone companies for use in connection with interstate and foreign message toll telephone service; and if
so

2. Whether, during the aforesaid period, such regulations and practices were unjust and unreasonable, and therefore unlawful within the
meaning of section 201(b) of the Communications Act of 1934, as amended,
or were unduly discriminatory or preferential in violation of section
202(a) of said Act.

The examiner found that there was a need and demand for a
device to connect the telephone landline system with mobile

radio systems which could be met in part by the Carterfone.
He also found that the Carterfone had no material adverse effect upon use of the telephone system. He construed the tariff
to prohibit attachment of the Carterfone whether or not it
harmed the telephone system, and determined that future prohibition of its use would be unjust and unreasonable. He also
found that it would be unduly discriminatory under section
202(a) of the Act, since the telephone companies permit the
use of their own interconnecting devices. However, he did not
find the tariff prohibitions to have been unlawful in the past,
largely because the harmless nature of the Carterfone was not
known to the telephone companies, and he did not find that a
general prohibition against non-telephone company supplied interconnecting devices was unjust or unwise, because of the risk
he saw of "serious harm to the heart of the nation's communications network."

We agree with and adopt the examiner's findings that the
Carterfone fills a need and that it does not adversely affect
the telephone system. They are fully supported by the record.
We also agree that the tariff broadly prohibits the use of interconnection devices, including the Carterfone. Its provisions are clear as to this. Finally, in view of the above
findings, we hold, as did the examiner, that application of
the tariff to bar the Carterfone in the future would be unreasonable and unduly discriminatory. However, for the reasons to be given, we also conclude that the tariff has been
unreasonable, discriminatory, and unlawful in the past, and
that the provisions prohibiting the use of customer-provided
interconnecting devices should accordingly be striken.

We hold that the tariff is unreasonable in that it prohibits
the use of interconnecting devices which do not adversely affect the telephone system. See *Hush-A-Phone Corp.* v. *U.S.*, 99 U.S.
App. D.C. 190, 193, 238 F. 2d 266, 269 (D.C. Cir., 1956), holding
that a tariff prohibition of a customer supplied "foreign attachment" was "in unwarranted interference with the telephone
subscriber's right reasonably to use his telephone in ways
which are privately beneficial without being publicly detri-
mental." The principle of Hush-A-Phone is directly applicable
here, there being no material distinction between a foreign
attachment such as the Hush-A-Phone and an interconnection device such as the Carterfone, so far as the present problem is
concerned. Even if not compelled by the Hush-A-Phone decision,
our conclusion here is that a customer desiring to use an interconnecting device to improve the utility to him or both the
telephone system and a private radio system should be able to
do so, so long as the interconnection does not adversely affect
the telephone company's operations or the telephone system's
utility for others. A tariff which prevents this is unreasonable; it is also unduly discriminatory when, as here, the telephone company's own interconnecting equipment is approved for

use. The vice of the present tariff, here as in Hush-A-Phone,
is that it prohibits the use of harmless as well as harmful
devices.

A.T.&T. has urged that since the telephone companies have
the responsibility to establish, operate and improve the telephone system, they must have absolute control over the quality,
installation, and maintenance of all parts of the system in
order effectively to carry out that responsibility. Installation of unauthorized equipment, according to the telephone
companies, would have at least two negative results. First, it
would divide the responsibility for assuring that each part of
the system is able to function effectively and, second, it
would retard development of the system since the independent
equipment supplier would tend to resist changes which would
render his equipment obsolete.

There has been no adequate showing that nonharmful interconnection must be prohibited in order to permit the telephone
company to carry out its system responsibilities. The risk
feared by the examiner has not been demonstrated to be substantial, and no reason presents itself why it should be. No
one entity need provide all interconnection equipment for our
telephone system any more than a single source is needed to
supply the parts for a space probe. We are not holding that
the telephone companies may not prevent the use of devices
which actually cause harm, or that they may not set up reasonable standards to be met by interconnection devices. These
remedies are appropriate; we believe they are also adequate to
fully protect the system.

Nor can we assume that the telephone companies would be
hindered in improving telephone service by any tendency of the
manufacturers and users of interconnection devices to resist
change. The telephone companies would remain free to make improvements to the telephone system and could reflect any such
improvements in reasonable revised standards for nontelephone
company provided devices used in connection with the system.
Manufacturers and sellers of such devices would then have the
responsibility of offering for sale or use only such equipment
as would be in compliance with such revised standards. An
owner or user of a device which failed to meet reasonable revised standards for such devices, would either have to have
the device rebuilt to comply with the revised standards or
discontinue its use. Such is the risk inherent in the private
ownership of any equipment to be used in connection with the
telephone system.

* * *

Accordingly, we find that tariff FCC No. 263, paragraphs
2.6.1 and 2.6.9 are, and have since their inception been, unreasonable, unlawful and unreasonably discriminatory under

sections 201(b) and 202(a) of the Communications Act of 1934, as amended.

* * *

BEFORE THE
FEDERAL COMMUNICATIONS COMMISSION
WASHINGTON, D.C. 20554

In Re Applications of	*Docket No. 16509*
Microwave Communications, Inc.	*File No. 4615-Cl-*
For Construction Permits To Establish	*P-64*
New Facilities in the Domestic Public	*Dockets Nos. 16510,*
Point-to-Point Microwave Radio Serv-	*16511, 16512, 16513,*
ice at Chicago, Ill., St. Louis, Mo.,	*16514, 16515, 16516,*
and Intermediate Points	*16517, 16518, 16519*

DECISION
(Adopted August 13, 1969)

COMMISSIONER BARTLEY FOR THE COMMISSION: CHAIRMAN HYDE DIS-
SENTING AND ISSUING A STATEMENT; COMMISSIONER ROBERT E.
LEE DISSENTING AND ISSUING A STATEMENT IN WHICH COMMIS-
SIONER WADSWORTH JOINS; COMMISSIONER JOHNSON CONCURRING
AND ISSUING A SEPARATE STATEMENT.

1. This proceeding involves applications filed by Microwave
Communications, Inc. (MCI), for construction permits for new
facilities in the Domestic Public Point-to-Point Radio Service
at Chicago, Ill., St. Louis, Mo., and nine intermediate points.
MCI proposes to offer its subscribers a limited common carrier
microwave radio service, designed to meet the interoffice and
interplant communications needs of small business. Its subscribers would be able to lease microwave channels in varying
bandwidths in increments of 2 kc. for either the entire length
of its system or any segment thereof. For broadband users,
channels may be leased in increments of 48, 250, and 1,000 kc.
MCI, however, does not plan to provide its subscribers with a
complete microwave service. The proposed service would be
limited to transmission between MCI's microwave sites, making
it incumbent upon each subscriber to supply his own communications link between MCI's sites and his place of business (loop
service).

2. MCI contends that it will offer its subscribers substantially lower rates than those charged for similar services by
the established carriers and that subscribers with less than
full-time communication needs will be able to achieve additional savings through the channel sharing and half-time use
provisions of its proposed tariff. Up to five subscribers will
be permitted to share each channel on a party-line basis with a
pro-rate reduction in rates. MCI will lease channels for half-
time use between 6 a.m. and 6 p.m. with a 25-percent reduction
in rates; and between 6 p.m. and 6 a.m. it proposes to combine
the channels leased for half-time use into broadband channels
of 48, 250, or 1,000 kc. for high-speed data transmission. MCI

further asserts that its proposed tariff contains fewer restrictions than those of the existing common carriers, so that
greater flexibility of use will be possible, particularly with
respect to channel bandwidth, splitting channels for voice and
data transmissions, and in the attachment of customer equipment.

3. MCI's applications are opposed by Western Union Telegraph
Co. (Western Union), General Telephone Co. of Illinois (General), and the Associated Bell System Cos., American Telephone &
Telegraph Co., Illinois Bell Telephone Co., and Southwestern
Bell Telephone Co. (Bell), which presently provide microwave
services to the geographical area which MCI proposes to serve.
In a memorandum opinion and order, F.C.C. 66-89, released February 8, 1966, and published in the Federal Register on February 11, 1966 (31 F.R. 2666), we designated the MCI applications for hearing on issues to determine inter alia: (a)
whether the established common carriers offer services meeting
the needs which MCI proposes to meet in the area which MCI proposes to serve; (b) whether the grant of MCI's applications
would result in wasteful duplication of facilities; (c) whether
MCI is financially qualified to construct and operate its proposed facilities; (d) whether there is a need for MCI's proposal; and (e) whether operation of MCI's proposed system would
result in interference to existing common carrier services.

* * *

5. Upon release of the initial decision, we recognized that
the questions raised at the hearing involved important policy
considerations respecting the entry of new licensees into the
communications common carrier field; and we directed by order,
F.C.C. 67-1244, adopted November 14, 1967, that the initial
decision be referred to the Commission for review. Bell, General, and Western Union each filed exceptions to the initial
decision and requested oral argument. On April 30, 1968, we
heard oral argument, en banc. We have considered the initial
decision in light of the record, pleadings, and oral argument.
Except as modified below and in the attached appendix, we adopt
the hearing examiner's findings and conclusions.

6. The principal contentions advanced by Bell, General, and
Western Union against the grant of MCI's applications are: (1)
That MCI is not financially qualified to construct and operate
the proposed facilities; (2) that no need has been shown for
the common carrier services proposed; (3) that MCI will be unable to provide a reliable communications service; (4) that
the proposal represents an inefficient utilization of the frequency spectrum; and (5) that the proposal is not technically
feasible. Each of these contentions will be considered below.

MCI's Financial Qualifications

* * *

NEED FOR THE SERVICE

22. MCI is offering a service intended primarily for interplant and interoffice communications with unique and special-

ized characteristics. In these circumstances we cannot per-
ceive how a grant of the authorizations requested would pose
any serious threat to the established carriers' price averaging
policies. Lower rates for the service offered is not the sole
basis for our determination that MCI has demonstrated a need
for the proposed facilities, but the flexibility available to
subscribers, and the sharing and the part-time features of the
proposal have been considered to be significant factors as well.
The case of WADS, 35 F.C.C. 149 (1963), cited by Bell is therefore inapposite. Here the potential demand for the new service
is not generated solely by reason of lower rates for a like
service but because there is a "need for service which, if not
met, would result in a serious deficiency in the communication
services available to the public" (35 F.C.C. at 155). It may
be, as the telephone companies and Western Union argue, that
some business will be diverted from the existing carriers upon
the grant of MCI's applications, but that fact provides no
sufficient basis for depriving a segment of the public of the
benefits of a new and different service.

23. Moreover, if we were to follow the carriers' reasoning
and specify as a prerequisite to the establishment of a new
common carrier service that it be so widespread as to permit
cost averaging, we would in effect restrict the entry of new
licensees into the common carrier field to a few large companies which are capable of serving the entire Nation. Such
an approach is both unrealistic and inconsistent with the public interest. Innovations in the types and character of communications services offered or economics in operation which
could not at once be instituted on a nationwide basis would be
precluded from ever being introduced. In the circumstances of
this case, we find the cream skimming argument to be without
merit.

                              * * *

26. The conclusion that a public need exists for the services proposed by MCI is adequately supported by the evidence
of record and we have relied solely upon that evidence in
reaching our determination. Consequently, it is unnecessary
to take official notice of A.T.&T.'s aforementioned statements
or to place any reliance thereon in connection with the resolution of the public need issue; and we shall not do so.
Nevertheless, we cannot ignore statements made by a party in
filings with the Commission which contradict or are inconsistent with the position taken by that party in an adjudicatory
proceeding. The statements by A.T.&T. in support of its own
proposals substantially undermine the arguments advanced by
the carrier in this proceeding to the effect that no public
need exists for the sharing provisions of the MCI proposal and
that MCI is cream-skimming. To the extent necessary to demonstrate the different positions taken by A.T.&T., we shall take
official note of the statements made in support of its proposed
tariff revisions.

Reliability of Service
* * *

30. Despite the strong protests of the existing carriers we
are not persuaded that MCI's proposed equipment is technically
deficient in any material respect. The sheds to be used for
housing the equipment are waterproof and will afford sufficient
protection against adverse weather for the microwave transmitting and receiving equipment which has an operational range
from -30° to +50° C. While working conditions in the sheds in
case of a breakdown would not be of the best, there is nothing
in the record to justify a finding that they are not usable.
It may be, as the carriers contend, that reliance upon gener-
ators instead of batteries is preferable. Nevertheless, it
appears, that batteries will keep the system operational during
periods of all but exceptionally extended power outages; and in
that event the batteries will suffice until another power
source is made available. We also find that MCI's arrangements
for maintenance are adequate. By reason of the relatively
limited size of its system, MCI will not employ a full-time
maintenance staff and it would be uneconomical to do so. In
lieu thereof, MCI has contracted with two firms, both of which
are experienced in installing and maintaining private microwave systems, to periodically inspect the microwave facilities
and to perform all necessary repair and maintenance work. Each
microwave relay site will be equipped with a fault alarm connected to a 24-hour answering service which, in turn, will
contact the appropriate maintenance company. These companies
operate on a 24-hour basis and no tower site is located more
than 106 miles from one of them. On the basis of this showing,
we are satisfied that the steps proposed to be taken by MCI for
maintenance and repair of its equipment should keep to a minimum any interruption in service resulting from equipment or
power failure.

31. Western Union's criticism that the proposed system is
deficient in that no provision is made for alternate routing
in the event of damage or outage at any point on the system
must be rejected as without merit. The alternate routes maintained by the major carriers serve also as primary routes for
other cities and thus have a dual purpose. However, were MCI
to construct an alternate route it would lie fallow except
during periods of extended outages and such limited use would
constitute a waste of frequencies which would not be in the
public interest. MCI has available substitute towers and other
equipment. It is prepared in case of outages to make repairs
expeditiously and, with the possible exception of a major
catastrophe, to resume operations with a minimum of delay.

Efficient Utilization of the Frequency Spectrum

32. We recognize, as the carriers argue, that MCI will not
make the fullest possible use of the frequencies which it

seeks. MCI intends to use frequency diversity in the operation of its system, providing one protection channel for one
working channel. Each MCI channel will have a maximum capacity of 960 circuits, but MCI anticipates an initial loading
of between 45 and 75 circuits per channel and a maximum load
of 300 circuits. In contrast, Bell uses two protection channels for six working channels, and it claims that each of its
channels has a maximum capacity of 1,800 circuits. Even if
Bell's systems do not achieve maximum loading, it clearly
transmits a greater number of message circuits per channel
than that contemplated by MCI.

33. In view of the limited frequencies available for common carrier use, the increasing demand for such frequencies,
and the possibility that the grant of an application might
limit a future assignment to a carrier which proposes a heavier loading or a better service, efficient utilization of the
common carrier frequency spectrum by an applicant is a matter
of serious concern to the Commission. In arriving at a public-
interest determination in this case, we have therefore given
careful consideration to the merits and demerits of MCI's proposed use of frequency diversity, its light loading, and other
factors which go to efficient utilization of the common carrier
portion of the frequency spectrum. With respect to MCI's proposed operation with frequency diversity, we note that the use
of one protection channel for one working channel complies with
existing policy. While the promulgation of rules to cover the
use of frequency diversity is presently under consideration,
adverse action on MCI's applications is not warranted for that
reason. We do not know what provisions ultimately will be
adopted and it may be that MCI's system will comply with the
new rules, or that, with some minor adjustments, it can be
brought into compliance.

34. Furthermore, in determining whether there is a frequency
wastage we must take into account the benefits to be derived by
the public from the proposed common carrier facilities, and to
weigh these benefits against the disadvantages alleged by the
opposing carriers. We have found that by reason of its low-
cost, sharing, and part-time use provisions, MCI can reasonably
be expected to furnish an economical microwave communications
service to a segment of the public which presently cannot avail
itself of such a service; and that its flexibility features
will enable potential users to make more efficient use of their
business equipment. These are substantial benefits which, in
our view, outweigh the fact that MCI will not make the fullest
possible use of its frequencies. When frequencies are used to
meet a significant unfulfilled communications need, we do not
believe that such use may be considered as "inefficient."

The Feasibility of Loop Service

35. The testimony of MCI's public witnesses and the findings of the Spindletop survey show that, in general, MCI's

potential subscribers have no interest in providing their own
communications link between their facilities and MCI's transmitter sites. Therefore, MCI's ability to market its service
will be dependent on the ability of its subscribers to secure
loop service from the other common carriers serving the service area.
   36. We are not unmindful of the fact that the carriers maintain that loop service is not technically feasible and that
there is no provision for such service in their tariffs. However, insufficient evidence is contained in this record to
support a conclusion that the proposed interconnection is not
feasible, and we are not disposed to deny MCI's applications
on the basis of the unsubstantiated allegations which have been
advanced herein by the telephone companies and Western Union.
What seems a more likely obstacle to interconnection is, as
the hearing examiner indicated, the "carriers' intransigence,
manifested in this case***." In these circumstances, the carriers are not in a position to argue that consideration of the
interconnection question is premature. Since they have indicated that they will not voluntarily provide loop service we
shall retain jurisdiction of this proceeding in order to enable
MCI to obtain from the Commission a prompt determination on the
matter of interconnection. Thus, at such time as MCI has customers and the facts and details of the customer's requirements
are known, MCI may come directly to the Commission with a request for an order of interconnection. We have already concluded that a grant of MCI's proposal is in the public interest. We likewise conclude that, absent a significant showing
that interconnection is not technically feasible, the issuance
of an order requiring the existing carriers to provide loop
service is in the public interest.

Summary
   37. This is a very close case and one which presents exceptionally difficult questions. We have found MCI to be
financially qualified but we realize that any unforeseen circumstances requiring a sizable expenditure may impair the
applicant's financial capacity. We have found, based on the
weight of the evidence, that there is a substantial likelihood
that the communications of its subscribers will arrive promptly,
in accurate form, and without extended interruptions due to
failures in the system. We wish to make clear, therefore, that
the findings and conclusions reached herein apply only to the
frequencies specified, and for the areas described, in the applications now pending before us. Should MCI seek to obtain
additional frequencies or to extend its microwave service to
new areas, our action on its application will be based on a
close scrutiny of its operations, the rules then governing the
grant of applications for common carrier microwave frequencies
and all other applicable policy considerations. Likewise, in
connection with an application for renewal of license, we may

deny the application if circumstances so warrant or grant renewal on such conditions as we deem essential to insure that
MCI's subscribers receive a reliable transmission service of
acceptable quality. However, it would be inconsistent with
the public interest to deny MCI's applications and thus deprive the applicant of an opportunity to demonstrate that its
proposed microwave facilities will bring to its subscribers
the substantial benefits which it predicts and which we have
found to be supported by the evidence in this proceeding. We
conclude, on the basis of the record as a whole, that the public interest will be served by a grant of MCI's application.

* * *

The Bell Bill
This is the wording of a proposed statute introduced in 1977
defining a policy to be followed regarding competition in
the common carrier area of telecommunications. AT&T proposed
the statute which clearly is a reaction to the MCI competitive
threat.

95th Congress
1st Session
H. R. 8
IN THE HOUSE OF REPRESENTATIVES
January 4, 1977
A BILL

To reaffirm the intent of Congress with respect to the structure of the common carrier telecommunications industry render-
ing services in interstate and foreign commerce; to grant additional authority to the Federal Communications Commission to
authorize mergers of carriers when deemed to be in the public
interest; to reaffirm the authority of the States to regulate
terminal and station equipment used for telephone exchange
service; to require the Federal Communications Commission to
make certain findings in connection with Commission actions
authorizing specialized carriers; and for other purposes.

Be it enacted by the Senate and House of Representatives
of the United States of America in Congress assembled, That
this Act may be cited as the "Consumer Communications Reform
Act of 1977".

CONGRESSIONAL FINDINGS AND DECLARATION OF PURPOSE
SEC. 2. The Congress finds and declares that--
(a) The revenues from integrated interstate and foreign
common carrier telecommunications services, based on charges
reflecting both costs and value of service, have contributed
toward meeting the costs of facilities used in common for providing such interstate and foreign services and local telephone
exchange service throughout the United States, and thereby
helped maintain a level of charges for telephone exchange service which is lower than otherwise would be required.

(b) The technical integrity of the nationwide telecommunications system, its coordinated planning, design, installation, improvement, management, operation and maintenance are indispensable elements in the interstate telecommunications network, necessary both to the reasonableness of charges and to the high quality and universality of common carrier telecommunications service, and accordingly Congress hereby reaffirms its policy that the integrated interstate telecommunications network shall be structured so as to assure widely available, high quality telecommunications services to all of the Nation's telecommunications users.

(c) The authorization of lines, facilities, or services of specialized carriers which duplicate the lines, facilities, or services of other telecommunications common carriers--

(1) involves higher charges for users of telephone exchange service by decreasing the interstate revenues that otherwise would be available for contribution to the common costs of providing telephone services throughout the United States;

(2) fosters inefficiencies in the utilization of national telecommunications resources through the creation of unnecessary and wasteful duplication of telecommunications lines and facilities and wasteful use of the radio spectrum;

(3) significantly impairs the technical integrity, the coordinated planning, design, installation, improvement, management, operation and maintenance of the integrated nationwide telecommunications network; and

(4) has an adverse impact on the national objectives of maintaining stability of consumer price levels, conserving national economic resources, improving productivity, and fostering an economy that will maintain adequate sources and reasonable costs of capital;

and is, therefore, contrary to the public interest.

(d) The Congress reaffirms its intent that the complete authority to regulate terminal and station equipment used for telephone exchange service shall rest with the States even though such terminal and station equipment also may be used in connection with interstate services.

(e) The congressional findings and declarations of policy set forth herein are necessary to achieve the purposes of the Communications Act of 1934 as specified in section 1 of that Act; and the Federal Communications Commission shall take no action inconsistent with the findings and declarations in this Act.

CHARGES FOR SERVICE

SEC. 3. Section 201(b) of the Communications Act of 1934, as amended (47 U.S.C. 201) is amended by adding the following at the end of the first sentence: "No compensatory charges for or in connection with such communication service may be found

to be unjust or unreasonable on the ground that it is too low.
The Commission may not hold the charge of a carrier up to a
particular level to protect the traffic or revenues from a communication service offered or provided by another carrier if
such charge proposed by the carrier is compensatory. As used
in this subsection, a charge is compensatory so long as it
equals or exceeds the incremental cost of providing the communications service. Such incremental cost is the additional
cost caused by the provision of the service including, where
appropriate, the capital costs of whatever additional facilities are required to provide the service.".

ACQUISITIONS BY AND OF CERTAIN COMMON CARRIERS
SEC. 4. The Communications Act of 1934, as amended, is
further amended by adding the following new section 224:
"SEC. 224. Upon application of any common carrier or other
person involved in the transaction, the Commission shall have
jurisdiction (i) to approve the acquisition of control by a
domestic common carrier of any other domestic common carrier
or the acquisition of the whole or any part of the property of
a domestic common carrier by any other domestic common carrier,
or (ii) to approve the acquisition by a person which is not a
common carrier of control of any domestic common carrier or
the acquisition of the whole or any part of the property of a
domestic common carrier, whenever the Commission determines,
after full opportunity for hearing on an evidentiary record,
that such approval is in the public interest. The Commission
shall give reasonable notice in writing concerning any such
proposed action to the Governor of each of the States in which
the physical property affected, or any part thereof, is situ-
ated, and to each State commission that may also have jurisdiction over any of the common carriers involved, and to such
other persons as it may deem advisable, and shall afford such
parties a reasonable opportunity to participate in any hearings
related to such action. If the Commission approves the proposed acquisition, it shall certify to that effect; and there-
upon any Act or Acts of Congress making the proposed acquisition unlawful shall not apply. As used in this section 224,
'domestic common carrier' shall mean a common carrier, the
major portion of whose traffic and revenues is derived from
communications services other than foreign communications.
This section 224 shall not apply where either section 221(a)
or 222 of this Act is applicable or to the acquisition by any
person of a telephone common carrier as defined in section 225
(a)(1).".
SEC. 5. Section 2(b) of the Communications Act of 1934, as
amended, (47 U.S.C. 152(b)) is further amended by striking the
clause beginning with the words "except that" following the
semicolon and inserting the following "except that sections
201 through 205 of this Act, both inclusive, and section 224

of this Act shall, except as otherwise provided therein, apply to carriers described in clauses (2), (3), and (4).".

REAFFIRMATION OF STATE JURISDICTION OVER LOCAL
TERMINAL AND STATION EQUIPMENT

SEC. 6 Section 2(b) of the Communications Act of 1934, as amended (47 U.S.C. 152(b)) is further amended by striking "or" at the end of the phrase following "(1)" and substituting therefor the following: "including but not limited to, the charges, classifications, practices, services, facilities, or regulations for or in connection with the use or connection of any station equipment, terminating facilities, exchange plant, and other like instrumentalities and apparatus used in common for both intrastate communication service and interstate or foreign communication service, whether provided by a common carrier or any other person, or".

SEC. 7. Section 3 of the Communications Act of 1934, as amended (47 U.S.C. 153), is further amended by adding the following new subsection:

"(gg) 'Intrastate communication' means communication or transmission between points in the same State, territory, or possession of the United States, or in the District of Columbia, including among other things, all station equipment, terminating facilities, exchange plant, and other like instrumentalities and apparatus used for or in connection with telephone exchange service or interexchange service, even though such equipment, facilities, plant, instrumentalities or apparatus are or may be used in connection with interstate or foreign communications service. 'Intrastate communication service' means any service which provides intrastate communication.".

FINDINGS TO BE INCLUDED IN COMMISSION
AUTHORIZATIONS OF SPECIALIZED CARRIERS

SEC. 8 The following new section is added in title II of the Communications Act of 1934, as amended:

"SEC. 225. (a) As used in this section--

"(1) The term 'telephone common carrier' means any common carrier, the major portion of whose traffic and revenues, in interstate and foreign communication and in intrastate communication, is derived from message telephone services, telephone exchange services, radio-telephone exchange services, or a combination thereof.

"(2) The term 'telegraph common carrier' means any common carrier which provides a public message telegram service in interstate communications.

"(3) The term 'specialized carrier' means any common carrier other than a telephone or telegraph common carrier.

"(4) The term 'message telephone service' means telephone service between stations in different exchange areas on a message-by-message basis, contemplating a separate connection for each occasion of use.

"(5) The term 'public message telegram service' means a
substantially nationwide telegraph service for the transmission and reception of record matter where the transmission is not directly controlled by the sender and for which
a charge is collected on the basis of number of words transmitted and which is available to the public.

"(b) The Commission shall not grant or authorize any construction, acquisition, or operation of any communication or
transmission line or facility, or extension thereof, or any
modification or renewal thereof, that otherwise might be granted or authorized pursuant to any provision of this Act, to any
specialized carrier that furnishes or proposes to furnish interstate communication service unless the Commission shall
find, after full opportunity for evidentiary hearing on the
record, that such permit, license, or certificate, will not
result in increased charges for telephone exchange service or
in wasteful or unnecessary duplication of communication lines,
facilities, equipment and instrumentalities of any telephone
or telegraph common carrier, and will not significantly impair
the technical integrity and capacity for unified and coordinated planning, management, design, and operation of the nationwide telephone network. In finding that such grant or
authorization will not result in wasteful or unnecessary duplication, the Commission shall determine, among other things,
that the proposed service or services of the specialized carrier, which are the subject of the requested grant or authorization, (i) are not like or similar to any service or services
provided by a telephone or telegraph common carrier and (ii)
cannot be provided by available communications lines, facilities, equipment, or instrumentalities of a telephone or telegraph common carrier. At any hearing involving a matter under
this subsection, the burden of proof to support the requisite
findings by the Commission shall be on the applicant for such
permit, license, or certificate.".

References

Chapter One: Telecommunications Technology and Society: An Overview

(1) Report of the Surgeon General's Scientific Advisory Committee on Television and Social Behavior, (Washington, D.C.: U.S.G.P.O., 1971).
(2) P. Davison and F. Yu, eds., *Mass Communications Research: Major Issues and Future Directions*, (New York: Praeger, 1974).
(3) *Events in Telephone History*, (New York: AT&T, 1958) 13.
(4) R. Gable, "The Early Competitive Era in Telephone Communication, 1893-1920," 34 *Law and Contemporary Problems* 340 (1969).
(5) L. Waverman, "The Regulation of Intercity Telecommunications," in A. Phillips, ed., *Promoting Competition in Regulated Markets* (Washington, D.C., The Brookings Institution, 1975) 201-239.
(6) *In the Matter of Use of the Carterfone Device in Message Toll Service*, 13 F.C.C. 2d 420, (1958).
(7) *In re Applications of Microwave Communications, Inc.*, 18 F.C.C. 2d 953, (1969).
(8) E. Barnouw, *A Tower in Babel: A History of Broadcasting in the United States*, Vol. 1, (New York: Oxford University Press, 1966).
(9) *New York Times*, "TV Makes Debut in South Africa," (May 11, 1975) 8.
(10) J. Spatafora, "TV and Radio Teaching Tools," 40 *The Education Digest* 57 (November 1974).

Chapter Two: Telecommunications Technology

(1) J. Martin, *Future Developments in Communications*, Prentice-Hall, Inc., Englewood Cliffs, N.J., 1971.
(2) *Scientific American*, Special issue on Communications containing eleven invited papers by leading experts on

various aspects of communication. Vol. 227, No. 3, September 1972.

(3) C. E. Shannon, "The Mathematical Theory of Communication," *Bell System Technical Journal*, July and October 1948.

(4) J. Ward, "Present and Probable CATV Broadband-Communication Technology," in I. Pool, ed, *Talking Back* (Cambridge: MIT Press, 1973) 139.

(5) Prescott C. Mabon, *Mission Communications: the Story of Bell Laboratories*, Murray Hill, Bell Laboratories: 1974.

(6) P. Baran, "Broadband Interactive Communication Services to the Home: Part I Potential Market Demand," 23 *IEEE Trans. on Communications*, 5 (1975).

Chapter Three: Politics, Economics, and Telecommunications Technology

(1) D. Easton, *A Systems Analysis of Political Life*, (New York: John Wiley & Sons, Inc., 1965).

(2) R. Coase, "The Federal Communications Commission," 2 *Journal of Law and Economics* 1 (1959).

(3) R. Coase, "The Interdepartment Radio Advisory Committee," 5 *Journal of Law and Economics* 17 (1962).

(4) E. Krasnow and L. Longley, *The Politics of Broadcast Regulation*, (New York: St. Martin's Press, 1973).

(5) 47 U.S.C. §§ 151 et. seq. (1971).

(6) 47 C.F.R. Parts 31, 33, 42, 61, 62, 64, and 73 (1976).

(7) L. Kohlmeier, Jr., *The Regulators: Watchdog Agencies and the Public Interest*, (New York: Harper & Row, 1969).

(8) M. Nadel, *Corporations and Political Accountability*, (Lexington, Mass.: D. C. Heath, 1976).

(9) 47 U.S.C. §§ 205; 303 (1971).

(10) R. Fenno, *The Power of the Purse: Appropriations Politics in Congress*, (Boston: Little Brown & Co., 1966).

(11) J. Manley, *The Politics of Finance: The House Committee On Ways and Means*, (Boston: Little Brown & Co., 1970).

(12) E. Krasnow and H. Shooshin, III, "Congressional Oversight: The Ninety-Second Congress and the Federal Communications Commission," 10 *Haw. J. of Legislation* 297 (1973).

(13) T. Cronin and S. Greenberg, eds., *The Presidential Advisory System*, (New York: Harper & Row, 1969).

(14) R. Barrow, "OTP and the FCC: Role of the Presidency and the Independent Agency in Communications," 43 *U. Cincinnati L. Rev.* 291 (1974).

(15) R. Berner, *Constraints on the Regulatory Process: A Case Study of Regulation of Cable Television*, (Cambridge: Harvard University Program on Information Technologies and Public Policy. 1975).

(16) R. Wells and J. Grossman, "Concept of Judicial Policy-Making: A Critique," 15 *Journal of Public Law* 286 (1966).

(17) M. Shapiro, *Law and Politics in the Supreme Court*, (New York: Free Press of Glencoe, 1964).

(18) *Southwestern Cable Co. v. United States*, 392 U.S. 157 (1968).

(19) *Teleprompter Corp. v. Columbia Broadcasting System*, 415 U.S. 394 (1974).

(20) C. Vose, *Caucasians Only*, (Berkeley: University of California Press, 1959).

(21) A. Phillips, Jr., *Promoting Competition in Regulated Markets*, (Washington, D.C.: The Brookings Institution, 1975).

(22) *Hush-A-Phone Corp. et.al. v. U.S. and F.C.C.*, 238 F.2d 266, (D.C. Cir. 1956).

(23) H. Trebing, "Common Carrier Regulation--The Silent Crisis," 34 *Law and Contemporary Problems* 299 (1969).

(24) Senate Commerce Committee Print, "Appointments to the Regulatory Agencies: The Federal Communications Commission and the Federal Trade Commission," 94th Congress, 2nd Sess.; April 1976.

(25) D. Levine, "Regulating the Use of the Radio Spectrum," 5 *Stanford Journal of International Studies* 21 (1970).

(26) J. Kildow, *INTELSAT: Policy-Maker's Dilemma*, (Lexington, Mass: Lexington Books, 1973).

(27) C. Lindbloom, *The Policy-Making Process*, (Englewood Cliffs, N.J.: Prentice-Hall, 1968).

(28) R. Noll, M. Peck, and J. McGowan, *Economic Aspects of Television Regulation*, (Washington, D.C.: The Brookings Institution, 1973).

(29) E. Barnouw, *A Tower in Babel. A History of Broadcasting in the United States*, Vol. 1, (New York: Oxford University Press, 1966).

(30) H. Levin, *The Invisible Resource: Use and Regulation of the Radio Spectrum*, (Baltimore: The Johns Hopkins University Press, 1971).

(31) R. Posner, "Natural Monopoly and its Regulation," 21 *Stanford L. Rev.* 548 (1969).

(32) L. Waverman, "The Regulation of Intercity Telecommunications," in (21) 201-239.

Chapter Four: Broadcast Communications

(1) R. Noll, M. Peck, and J. McGowan, *Economic Aspects of Television Regulation,* (Washington, D.C.: The Brookings Institution, 1973).

(2) H. Levin, *The Invisible Resource: Use and Regulation of the Radio Spectrum,* (Baltimore: The Johns Hopkins Press, 1971).

(3) H. Levin, "The Radio Spectrum Resource," 11 *Journal of Law and Economics,* 433 (1968).

(4) R. Coase, "The Interdepartment Radio Advisory Committee," 5 *Journal of Law and Economics* 17 (1962).

(5) H. Levin, "New Technology and the Old Regulation in Radio Spectrum Management," 56 *Am. Econ. Rev. Papers and Pros.* 339 (1966).

(6) E. Krasnow and L. Longley, *The Politics of Broadcasting Regulation,* (New York: St. Martin's Press, 1973).

(7) L. Jaffe, "WHDH: The F.C.C. and Broadcasting License Renewal," 82 *Harvard Law Review* 1693 (1969).

(8) P.L. 87-445; 76 Stat. 64 (1962); 47 U.S.C. § 303(s) (1971).

(9) D. Webbink, "The Impact of UHF Promotions: The All-Channel Television Receiver Law," 34 *Law and Contemporary Problems* 535 (1969).

(10) 47 C.F.R. § 15.68(d)(3) (1976).

(11) *On the Cable: The Television of Abundance,* Report of the Sloan Commission on Cable Communications, (New York: McGraw-Hill, 1971).

(12) *Frontier Broadcasting Co. v. Laramie Community TV Co.,* 24 F.C.C. 251 (1958).

(13) *Carter Mountain Transmission Corp.,* 32 F.C.C. 459 (1962) aff'd. 321 F.2d 359 (D.C. Cir. 1963), cert den'd 375 U.S. 951 (1963).

(14) First Report and Order, 30 F.C.C. 683 (1969).

(15) Second Report and Order, 2 F.C.C. 2d 725 (1966).

(16) U.S. v. Southwestern Cable Company, 392 U.S. 157 (1968).

(17) Committee for Economic Development, *Broadcasting and Cable Television: Policies for Diversity and Change,* New York: CED, 1975) Appendix A.

(18) Third Report and Order, 36 F.C.C. 2d 141 (1972).

(19) B. Mitchell with R. Smiley, "Cables, Cities, and Copyrights," 5 *Bell Journal of Economics and Management Science* 235 (1974).

(20) R. Park, "Prospects for Cable in the 100 Largest Television Markets," 3 *Bell Journal of Economics and Management Science* 130 (1972).

(21) M. Seiden, *Cable Television U.S.A.: An Analysis of Government Policy,* (New York: Praeger, 1972).
(22) 47 C.F.R. § 76.251(a)(1) (1973).
(23) 47 C.F.R. § 76.252(b) (1976).
(24) 47 C.F.R. § 76.251(a)(3) (1973).
(25) 47 C.F.R. § 76.251(a)(4)-(6) (1973).
(26) 47 C.F.R. § 76.254(b) and (c) (1976).
(27) 47 C.F.R. § 76.251(a)(10)(i) and (ii) (1973).
(28) *Fortnightly Corp. v. United Artists Television Corp.,* 392 U.S. 390 (1968).
(29) *Teleprompter Corp. v. Columbia Broadcasting System,* 94 S. Ct. 1129 (1974).
(30) P.L. 94-553, 90 Stat. 2541 (1976), 17 U.S.C.A. §§ 101-810 (1976).
(31) 17 U.S.C.A. § 111 (1976).
(32) 17 U.S.C.A. § 801(b)(2) (1976).
(33) W. Baer and C. Pilmck, "Pay Television at the Crossroads," (Rand Corporation P-5159, April 1974).
(34) 47 C.F.R. § 76.225 (1976).
(35) C. Sterling, "Decade of Development: FM Radio in the 1960's," 48 *Journalism Quarterly* 222 (1971).
(36) S. 585, 93rd Cong., 2nd Sess., Passed Senate on June 13, 1974.
(37) F.C.C., 39th Annual Report, (Washington, D.C.: U.S.G.P.O., 1973).

Chapter Five: Two-Way Telecommunications

(1) M. Snitzer, "Re CB Pleasure Seekers," 5 *Popular Electronics* 6 (April, 1974).
(2) N. Johnson, "Towers of Babel: The Chaos in Radio Spectrum Utilization and Allocation," 34 *Law and Contemporary Problems* 505 (1969).
(3) 47 U.S.C. §§ 151 et. seq. (1971).
(4) R. Posner, "Natural Monopoly and Its Regulations," 21 *Stanford Law Review* 548 (1969).
(5) A. Kahn, *The Economics of Regulation: Principles and Institutions,* Vol. 2, (New York: John Wiley & Sons, 1971).
(6) F.C.C. Docket #20003.
(7) H. Averch and F. Johnson, "Behavior of the Firm Under Regulatory Constraint," 52 *American Economic Review* 1052 (1962).
(8) W. Sheperd, "The Competitive Margin in Communications" in W. Capron, ed., *Technological Change in Regulated Industries,* (Washington, D.C.: The Brookings Institution, 1971).

(9) A. Clarke, "Extra-Terrestial Relays," 51 *Wireless World*
     305 (1945).
(10) P. Bergellini, "The Intelsat Satellite Communications
     Network," 12 *Communications Society* 9 (November, 1974).
(11) J. Pierce, *The Beginnings of Satellite Communication,*
     (San Francisco: San Francisco Press, 1968).
(12) B. Edelson and L. Pollack, "Satellite Communications,"
     Science, Vol. 195, March 18, 1977, pp. 1125-33.
(13) Communications Satellite Act, 47 U.S.C. §§ 701 et. seq.
     (1971).
(14) W. Redisch, "ATS-6 Description," *Proc. International Com-
     munications Conference,* (San Francisco, June 1975) 18-1.
(15) A. Whalen, "Health Education Telecommunications Experi-
     ment," *Proc. International Communications Conference,*
     (San Francisco, June 1975) 18-6.
(16) J. Galloway, *The Politics and Technology of Satellite
     Communications,* (Lexington, Mass.: Lexington Books,
     1972).
(17) *Authorized Entities and Authorized Users Under the Com-
     munications Satellite Act of 1962,* 4 F.C.C. 2d 421,
     (1966).
(18) "Reports on Selected Topics in Telecommunications,"
     Published for National Academy of Engineering by National
     Academy of Sciences, (Washington, D.C. 1969).
(19) *Applications of ITT Cable and Radio Inc.,* 5 F.C.C. 2d 823
     (1966).
(20) M. Seiden, *Cable Television U.S.A.: An Analysis of Gov-
     ernment Policy* (New York: Praeger, 1972).
(21) *Final Report of the President's Task Force on Communica-
     tions Policy,* (Washington, D.C.: U.S.G.P.O. 1968) esp.
     Chapters 2 and 5.
(22) W. Melody, "Interconnection: Impact on Competition--
     Carriers and Regulation," in S. Winkler, ed., *Computer
     Communications: Impacts and Implications,* (Washington,
     D.C.: First International Conference on Computer Communi-
     cation, 1972) 445.
(23) *In the Matter of Hush-A-Phone Corp. et. al.,* 20 F.C.C.
     391 (1955); rev'd 238 F.2d 266 (D.C. Cir. 1956).
(24) *Hush-A-Phone Corp. et.al. v. U.S. and F.C.C.,* 238 F.2d
     266 (D.C. Cir. 1956).
(25) *In the Matter of Use of the Carterfone Device in Message
     Toll Service,* 13 F.C.C. 2d 420 (1968).
(26) *Statistical Abstract of the United States,* Table No. 826
     (Washington, D.C.: U.S.G.P.O., 1974).
(27) M. Irwin, "The Communication Industry and the Policy of
     Competition," 14 *Buffalo Law Review* 256 (1964).
(28) *In the Matter of Allocation of Frequencies in the Bands
     Above 890 MHz.,* 27 F.C.C. 359 (1959).

(29) K. Cox, "The Promise and Peril of Competition in Intercity Communications," in S. Winkler, ed., *Computer Communications: Impacts and Implications*, (Washington, D.C.:
First International Conference on Computer Communication,
1972).

(30) H. Trebing, "Common Carrier Regulation--The Silent
Crisis," 34 *Law and Contemporary Problems* 299 (1969).

(31) *In the Matter of AT&T Co., Tariff F.C.C. No. 250, TELPAK
Service and Channels*, 37 F.C.C. 1111 (1964).

(32) *In re Applications of Microwave Communications, Inc.*, 18
F.C.C. 2d 953 (1969).

(33) J. Martin, *Future Developments in Telecommunications*,
(Englewood Cliffs, N.J.: Prentice-Hall, 1971).

(34) *In the Matter of TELPAK Tariff Sharing Provisions of
AT&T and the Western Union Telegraph Co.*, Docket #17457.

(35) L. Waverman, "The Regulation of Intercity Telecommunica-
tions," in A. Phillips, ed. *Promoting Competition in
Regulated Markets*, (Washington, D.C.: The Brookings
Institution, 1975) 201.

(36) "Dutran Asks Court Review of DUV Authorization," 188
Telephony 9-10 (February 10, 1975).

(37) E. Dickson in association with R. Bowers, *The Video-
Telephone: Impact of a New Era in Telecommunications*,
(New York: Praeger, 1974).

(38) "The Picturephone System," Special Issue, 50 *Bell System
Technical Journal*, 219-709 (1971).

(39) L. Hardeman, "Picturephone to Change Its Image," 46
Electronics 75-76 (September 13, 1973).

(40) C. Epstein, "Traveling by Telephone--Get Ready for a New
Market," 188 *Telephony* 24-30 (April 28, 1975).

(41) J. Pierce, "New Trends in Electronic Communication," 63
American Scientist 31 (January, 1975).

(42) G. Gobl, "The Picture Telephone Makes Communication More
Efficient" (Cardiff: UWIST Lectures, September, 1974).

(43) P. McManamon, "Technical Implications of Teleconference
Service," *IEEE Trans. on Communications* 30 (1975).

(44) *Visual Telephone Service: Organizational Application of
the Picture Telephone for Improved Telecommunications
Management* (Fack 135 01, L.M. Ericsson, 1973).

(45) N. Johannesson, *Slow Scan Techniques and Devices*, (L.M.
Ericsson Mimeo Doc. No. Ue 4170, September, 1974).

(46) N. Johannesson and J. Reissen, *The Picture Telephone Network at L.M. Ericsson: Equipment, Applications and Implications* (L.M. Ericsson Mimeo Doc. No. Ue 4146, August,
1974).

(47) *Newsbriefs Background*, Western Electric Co., (May 5,
1975).

Chapter Six: Telecommunications' Impacts on Society

(1)   I. Pool and H. Alexander, "Politics in a Wired Nation," Report prepared for the Sloan Commission on Cable Communications (September, 1971).

(2)   G. Tuchman, ed., *The TV Establishment: Programming for Power and Profit* (Englewood Cliffs, N.J.: Prentice-Hall, 1974).

(3)   S. Tuchman and T. Coffin, "The Influence of Electron Night Television Broadcasts in a Close Election," 35 *Public Opinion Quarterly* 315 (1971).

(4)   A. Weston, ed., *Information Technology in a Democracy* (Cambridge: Harvard University Press, 1971).

(5)   J. Rule, *Private Lives and Public Surveillance* (London: Allen Lane, 1973).

(6)   R. Mendes, "Post-Industrialization and the Totalitarian State," Paper prepared for presentation at the Western Political Science Assn. Meeting, Portland, Ore., March, 1972).

(7)   M. Price, "The First Amendment and Television Broadcasting by Satellite," 23 *UCLA L. Rev.* 879 (1976).

(8)   J. Martin, *Future Developments in Telecommunications* (Englewood Cliffs, N.J.: Prentice-Hall, Inc., 1971).

(9)   J. Pierce, "On Some Social Aspects of Telecommunications," 13 *Communications Society* 10 (March, 1975).

(10) Report of the Surgeon General's Scientific Advisory Committee on Television and Social Behavior (Washington, D.C.: U.S.G.P.O., 1971).

(11) J. Spatafora, "TV and Radio Teaching Tools," 40 *The Education Digest* 57 (1974).

(12) G. Wamsley and R. Price, "Television Network News: Re-Thinking the Iceberg Problem" 25 *Western Political Quarterly* 434 (1972).

(13) Television and Adolescent Aggressiveness, A Technical Report to the Surgeon General's Scientific Advisory Committee on Television and Social Behavior, Vol. 3 (Washington, D.C.: U.S.G.P.O., 1971).

(14) S. Milgram and R. L. Shotland, *Television and Antisocial Behavior*, (New York: Academic Press, 1973).

(15) S. Feshback and R. Singer, *Television and Aggression: An Experimental Field Study* (San Francisco: Jossey-Bass, Inc., 1971).

Glossary

Allocation. The process of distributing any resource among various claimants. Allocation can be done by economic processes such as the market place or political processes involving lobbying.

AM. Amplitude modulation.

AM broadcasting. Radiating into free space a radio wave carrying information in the form of amplitude modulation of the carrier. Transmission bandwidth is twice that of the modulating signal bandwidth.

Amplitude modulation. Process whereby the amplitude of a radio wave is varied in accordance with the amplitude of a second signal such as speech, music, or television.

Analog signal. Signal whose amplitude varies continuously as opposed to a signal that has been quantized and whose amplitude can only take on values from a discrete set of possible amplitudes.

Attenuation. Decrease in magnitude of amplitude or power during transmission of a signal from one point to another.

Bandwidth. The frequency interval outside of which the power spectrum of a signal is less than some specified fraction of its value at a reference frequency (usually the reference frequency is that at which the spectrum is a maximum). Measured in Hertz.

Bit. An abbreviation of BInary digiT. (1) Corresponds to the occurrence of (either) one of the characters of a language employing two different characters such as 0, 1 or mark, space. (2) Unit of information content equal to that of a message having an a priori probability of one half.

Broadband. An adjective used to describe signals that extend over a wide portion of the spectrum. For example, TV signals.

Carrier. A wave having characteristics that can be varied from a reference value by modulation.

273

Channel. A transmission path over which communication signals are sent.

COMSAT. The Communications Satellite Corporation created by an Act of Congress in 1962 to develop and operate the United States' portion of an international communications satellite system. It became the operating arm of the INTELSAT system when it became operational in 1964.

Cross-bar system. A telephone line switching system using cross-bar switches.

Data-Under-Voice. Technique for providing high-speed transmission of information in digital form using the lower portion of the frequency band in existing microwave networks not normally employed for voice transmission.

Echo. Name of an early experimental earth satellite consisting of a large metallized, inflated sphere used as a passive reflector for communication signals.

Electromagnetic wave. A wave characterized by variations of electric and magnetic fields. Radio waves.

ESS-1. Electronic Switching System being installed by Bell System.

F.C.C. Federal Communications Commission was created in 1934 to regulate wire and electromagnetic communication in the United States. This agency has jurisdiction over radio and television broadcasting, common carriers, cable television, and other interstate and international uses of the spectrum.

Frequency. The number of periods per second of a periodic function. Measured in units of hertz (Hz)--one cycle per second; kilohertz (kHz)--1,000 cycles per second; megahertz (MHz)--1,000,000 cycles per second; gigahertz (GHz)-- 1,000,000,000 cycles per second.

FM broadcasting. Radiating into free space a radio wave carrying information in the form of frequency modulation of the carrier.

Frequency division multiplexing. Process by which each of several modulating waveforms modulates a separate subcarrier and the subcarriers are spaced in frequency so that no two subcarriers occupy the same frequency band. Permits simultaneous transmission of several signals over the same channel.

Frequency modulation. Process whereby the frequency of a radio wave is varied in accordance with the amplitude of a second signal such as speech, music, or a binary coded signal. Transmission bandwidth is larger than the modulating signal bandwidth by a factor of five for FM broadcasting.

Geosynchronous orbit. Satellite orbit (altitude of 42,000 km or 22,300 mi) in which the satellite revolves around the earth at the same rate as the earth rotates (once each 24 hours). A geosynchronous satellite appears stationary above the earth's surface at some point along the equator.

Guard space. A blank segment of the spectrum located between adjacent channels in order to minimize interference between them.

IC. Integrated circuit.

Integrated circuit. A combination of interconnected circuit elements inseparably associated on or within a continuous substrate.

INTELSAT. The international satellite consortium formed in 1964 and in 1972 on a permanent basis to own and use the international communications satellite system. Member nations own shares and vote in proportion to their usage of the communications system. Earth stations are owned by the individual user countries. INTELSAT I, II, III, IV are the names and generation of the international communications satellites used to provide the services to member countries.

ITT. International Telephone and Telegraph is an international common carrier supplying communications services by radio satellite and undersea cable on a worldwide basis.

Large scale integration. Process whereby highly complex integrated circuits are fabricated.

Legislative oversight. The process of congressional investigation of a particular agency or administrative process by committee hearings and budget examinations. Although legislation doesn't result from this process, it is a central element of congressional control and influence over the administration of various policies and programs.

LSI. Large scale integration.

Modulation. Process by which certain characteristics of a carrier are varied in accordance with a modulating signal.

Microwave. Refers to the frequency range of electromagnetic radiation in which the wavelength is less than 30 cm (1 ft); i.e., upwards of 1,000 MHz.

Microwave relay. Intercity communication trunk in which signals are transmitted between towers 40 to 50 km (25-31 mi) apart.

Monopoly profits. The profits which can be earned by a monopolist when there is no competition for a market and the sole producer can charge whatever price he chooses. Such profits are those which exceed the profits obtainable in a competitive market setting where price is determined by supply of and demand for a product.

Multiplexing. Combining two or more signals into a single wave
from which the individual signal can be recovered.

OTP. The Office of Telecommunications Policy was created by
President Nixon in 1970 to provide advice and high level
studies of current telecommunications policy issues.

PBX. Private branch exchange used by companies for internal
distribution of telephone calls.

PCM. Pulse code modulation.

Pixel. Abbreviation for picture element. Smallest resolution
element in a television picture.

Polarization diversity. Means of simultaneously transmitting
two different signals on the same carrier by using two
electromagnetic waves having their electric fields oriented at right angles to each other so they can be separated
before detection.

President's Task Force on Telecommunication Policy. A special
task force created by President Johnson in 1967 to examine
various issues of telecommunications such as satellites,
cable television, and domestic common carriers.

Pulse code modulation. A modulation process whereby a waveform is first converted from analog to digital form and
the digital representation of the samples transmitted by
a binary code.

Quantization. Process by which a continuous range of amplitudes is divided into a nonoverlapping set of subranges
and to each subrange a unique value of the output is assigned.

RCA. Radio Corporation of America is an international common
carrier. It is also engaged in domestic satellite programs, broadcast network operations, and the manufacture
of electronic equipment.

RELAY. Early experimental communications satellite in low
earth orbit (1962-64).

Sampling. Process of obtaining a sequence of instantaneous
values of a waveform at regular intervals.

Signal-to-noise ratio. Relative power of the signal to that
of the noise in a communication channel. High signal-to-
noise ratio gives good transmission performance.

Single sideband transmission. A method of operation in which
one of the two sidebands produced by amplitude modulation
is suppressed. Transmission bandwidth equals modulating
signal bandwidth.

Solid state. Adjective used to describe a device, circuit, or
system whose operation depends on electrical phenomena
occurring in a solid such as crystals of semiconductor
materials.

Spectrum. The distribution of the components of a wave as a function of frequency. The range of components within which a signal from a particular kind of source lie; for example, audio frequency spectrum.

Telephony. The process of communicating by telephone.

Teletype. Trademark of Teletype Corporation referring to family of teleprinter equipment.

Telex. A dial-up telegraph service enabling subscribers to communicate directly with each other using teletypewriters.

Telpac. A rate structure or tariff arrangement created by AT&T in the early 1960's to provide competitive nonvoice transmission services. These rates have been the subject of continual controversy with AT&T competitors who challenge them as predatory and not reflective of real costs.

TELSTAR. Early experimental communications satellite in low earth orbit (1962-63).

Time division multiplexing. Process in which each of several modulating waveforms modulates a separate pulse subcarrier, the pulse subcarriers being spaced in time so that no pulses overlap any others. Permits simultaneous transmission of several signals over the same channel.

Transistor. An active semiconductor device with three or more terminals that is used for amplification and signal generation in electronic circuits.

Trunk. A circuit connecting two telephone exchanges in different localities. Also called a toll circuit.

TWX. Teletypewriter Exchange Service operated by AT&T whereby subscribers can communicate with each other by means of teletypewriters.

UHF. Ultra high frequency radio waves lying in the range of 300 to 3,000 MHz.

VHF. Very high frequency radio waves lying in the range of 30 to 300 MHz.

WATS. Wide Area Telephone Service provided by telephone companies that permits customers to make calls in a specific zone on a dial basis for a flat monthly charge.

Waveform. Variation in the amplitude of a signal as a function of time.

Index

Vested interests, 13, 61, 70-
  72, 95
VHF, TV, 34, 104-11 *passim*
Video telephone, 34, 176-82
Violence-oriented TV programs,
  201-2
Voice channels, 42-43, 45-56

WATS, 171, 187
Wave guide: 44-45, 91; circular, being installed, fig.,
  45

WESTAR, 158
Western Electric, 70, 148, 196
Western Union, 71, 158, 171
Wide Area Telephone System.
  See WATS
Wired city, 50